Shahid Ali Khan, Abdur Rauf, Guibin Xu (Eds.)
Hydrogels

Also of interest

Mechanics of Cells and Tissues in Diseases.
Volume 1: Biomedical Methods
Lekka, Navajas, Radmacher, Podestà (Eds.), 2023
ISBN 978-3-11-064059-5, e-ISBN (PDF) 978-3-11-064063-2,

Mechanics of Cells and Tissues in Diseases.
Volume 2: Biomedical Applications
Lekka, Navajas, Radmacher, Podestà (Eds.), 2023
ISBN 978-3-11-099972-3, e-ISBN (PDF) 978-3-11-098938-0,

Biofabrication
Forget, 2022
ISBN 978-1-5015-2335-9, e-ISBN (PDF) 978-1-5015-1573-6,

Corrosion Prevention Nanoscience.
Nanoengineering Materials and Technologies
Tukhlievich, Verma (Eds.), 2023
ISBN 978-3-11-107009-4, e-ISBN (PDF) 978-3-11-107175-6,

Biopolymers.
Environmental Applications
Aravind, Kamaraj (Eds.), 2023
ISBN 978-3-11-099872-6, e-ISBN (PDF) 978-3-11-098718-8,

Hydrogels

Antimicrobial Characteristics, Tissue Engineering, Drug
Delivery Vehicle

Edited by
Shahid Ali Khan, Abdur Rauf and Guibin Xu

DE GRUYTER

Editors
Dr. Shahid Ali Khan
Department of Chemistry
School of Natural Sciences
National University of Sciences and Technology
44000 Islamabad
Pakistan
Department of Urology
Key Laboratory of Biological Targeting Diagnosis
Therapy and Rehabilitation of Guangdong Higher Education Institutes
The Fifth Affiliated Hospital of Guangzhou Medical University
Guangzhou Medical University
Guangzhou 510700
China

Dr. Abdur Rauf
Department of Chemistry
University of Swabi
Anbar 23561
Pakistan

Prof. Dr. Guibin Xu
Department of Urology
Key Laboratory of Biological Targeting Diagnosis
Therapy and Rehabilitation of Guangdong Higher Education Institutes
The Fifth Affiliated Hospital of Guangzhou Medical University
Guangzhou Medical University
Guangzhou 510700
China

ISBN 978-3-11-133349-6
e-ISBN (PDF) 978-3-11-133408-0
e-ISBN (EPUB) 978-3-11-133422-6

Library of Congress Control Number: 2023944028

Bibliographic information published by the Deutsche Nationalbibliothek
The Deutsche Nationalbibliothek lists this publication in the Deutsche Nationalbibliografie;
detailed bibliographic data are available on the internet at http://dnb.dnb.de.

© 2024 Walter de Gruyter GmbH, Berlin/Boston
Cover image: Aiman Dairabaeva/iStock/Getty Images Plus
Typesetting: Integra Software Services Pvt. Ltd.
Printing and binding: CPI books GmbH, Leck

www.degruyter.com

Preface

Hydrogels are superabsorbent materials that are used in many technological applications. Researchers are trying to discover therapeutic agents that manage and cure the disease. With the advancement of science and technology, medicinal chemistry and material science synergetic efforts marketed many therapeutic agents for biomedical applications. Among various therapeutic agents, hydrogels are one of the most widely applied materials used to treat various diseases and one of the most diverse materials used for multipurpose applications. Hydrogels were the first biomaterials used for human beings. Hydrogels are polymeric networks and are water-insoluble; however, sometimes they are established as a colloidal gel in water. Hydrogels are superabsorbent materials because they can absorb more than 90% water, hence regarded as natural living tissues. Mechanically strong hydrogels were synthesized by the advent of new synthetic strategies. Owing to the swollen properties, three-dimensional polymer network, and strong mechanical characteristics, these are widely used in catalysis, adsorption, drug delivery systems for proteins, contact lenses, wound dressings, wound healing, bone regeneration, tissue engineering, baby diapers, food rheology, and many others. Due to their diverse applications, hydrogels are considered one of the smartest materials in pharmaceutics. As they are very sensitive to target, it is considered the favorite and preferred choice in biomedical sectors. Patients are psychologically scared of surgeries regarding huge expenses and failure. So, researchers are working on hydrogels as an alternative surgical replacement. In most cases, they have successfully achieved research on hydrogels in bone and tissue repair. It might be the hope of life for serious patients in the future. The domain of this work will cover state-of-the-art potentials, characteristics, properties, and applications in various technological areas.

Editors
Dr. Shahid Ali Khan
Department of Chemistry, School of Natural Sciences, National University of Science and Technology (NUST), Islamabad 44000, Pakistan; Department of Urology, Key Laboratory of Biological Targeting Diagnosis, Therapy and Rehabilitation of Guangdong Higher Education Institutes, The Fifth Affiliated Hospital of Guangzhou Medical University, Guangzhou Medical University, Guangzhou 510700, China
Dr. Abd ur Rauf
Department of Chemistry, University of Swabi, Anbar 23561, Pakistan
Prof. Dr. Guibin Xu
Department of Urology, Key Laboratory of Biological Targeting Diagnosis, Therapy and Rehabilitation of Guangdong Higher Education Institutes, The Fifth Affiliated Hospital of Guangzhou Medical University, Guangzhou Medical University, Guangzhou 510700, China

https://doi.org/10.1515/9783111334080-202

Contents

Contents

List of contributors

Chapter 1
Humaira Bibi
Department of Chemistry
Govt. Postgraduate College
Nowshera 24100
Khyber-Pakhtunkhwa
Pakistan

Waseeq Ur Rehman
Department of Chemistry
Govt. Postgraduate College
Nowshera 24100
Khyber-Pakhtunkhwa
Pakistan
and
Department of Chemistry
Bacha Khan University
Charsadda 24420
Khyber Pakhtunkhwa
Pakistan
gwaseeq@gmail.com

Shah Hussain
Department of Chemistry
Govt. Postgraduate College
Nowshera 24100
Khyber-Pakhtunkhwa
Pakistan

Kausar Shaheen
Department of Physics
University of Peshawar
Peshawar 25120
Khyber Pakhtunkhwa
Pakistan

Zarbad Shah
Bacha Khan University
Charsadda 24420
Khyber Pakhtunkhwa
Pakistan

Muhammad Asim
Department of Chemistry
Govt. Postgraduate College
Nowshera 24100
Khyber-Pakhtunkhwa
Pakistan

Shahid Ali Khan
School of Natural Sciences
National University of Science and Technology
Islamabad
Pakistan
and
Department of Urology
Key Laboratory of Biological Targeting Diagnosis
Therapy and Rehabilitation of Guangdong
Higher Education Institutes
The Fifth Affiliated Hospital of Guangzhou
Medical University
Guangzhou Medical University
Guangzhou 510700
China
shahid.ali@sns.nust.edu.pk

Chapter 2
Laiba Maryam
Department of Chemistry
School of Natural Sciences (SNS)
National University of Sciences and Technology
(NUST)
Islamabad, 44000
Pakistan

Asma Gulzar
Department of Chemistry
School of Natural Sciences (SNS)
National University of Sciences and Technology
(NUST)
Islamabad 44000
Pakistan

https://doi.org/10.1515/9783111334080-204

Dr. Mudassir Iqbal
Department of Chemistry
School of Natural Sciences (SNS)
National University of Sciences and Technology
(NUST)
Islamabad 44000
Pakistan
mudassir.iqbal@sns.nust.edu.pk

Chapter 3
Tahseen Arshad
Department of Chemistry
School of Natural Sciences (SNS)
National University of Sciences and Technology
(NUST)
Islamabad 44000
Pakistan

Muhammad Pervaiz
Department of Basic and Applied Chemistry
Faculty of Science and Technology
University of Central Punjab
Lahore 54000
Pakistan

Zhiduan Cai
Department of Urology
Key Laboratory of Biological Targeting Diagnosis
Therapy and Rehabilitation of Guangdong
Higher Education Institutes
The Fifth Affiliated Hospital of Guangzhou
Medical University
Guangzhou Medical University
Guangzhou 510700
China

Guibin Xu
Department of Urology
Key Laboratory of Biological Targeting Diagnosis
Therapy and Rehabilitation of Guangdong
Higher Education Institutes
The Fifth Affiliated Hospital of Guangzhou
Medical University
Guangzhou Medical University
Guangzhou 510700
China

Shahid Ali Khan
Department of Chemistry
School of Natural Sciences (SNS)
National University of Sciences and Technology
(NUST)
Islamabad 44000
Pakistan
and
Department of Urology
Key Laboratory of Biological Targeting Diagnosis
Therapy and Rehabilitation of Guangdong
Higher Education Institutes
The Fifth Affiliated Hospital of Guangzhou
Medical University
Guangzhou Medical University
Guangzhou 510700
China
shahid.ali@sns.nust.edu.pk

Chapter 4
Saurabh Shekhar
Department of Pharmacy
School of Medical and Allied Sciences
GD Goenka University
Sohna Gurgaon Road
Sohna, Haryana
India

Shailendra Bhatt
Department of Pharmacy
School of Medical and Allied Sciences
GD Goenka University
Sohna Gurgaon Road
Sohna, Haryana
India

Rohit Dutt
Gandhi Memorial National PG College
Ambala Cantt
Haryana
India

Manish Kumar
MM College of Pharmacy
MM (Deemed to be University)
Mullana-Ambala, Haryana
India

Rupesh K. Gautam
Department of Pharmacology
Indore Institute of Pharmacy
Rau, Indore
Madhya Pradesh
India
drrupeshgautam@gmail.com

Chapter 5
Shehla Khan
Department of Biotechnology
University of Swabi
Swabi
Anbar, KPK
Pakistan

Abdur Rauf
Department of Chemistry
University of Swabi
Anbar 23561
Pakistan

Chapter 6
Farzana Nazira
Department of Chemistry
School of Natural Sciences
National University of Sciences and Technology
(NUST)
Islamabad 44000
Pakistan
farzananazir88@gmail.com

Khadija Munawara
Department of Chemistry
School of Natural Sciences
National University of Sciences and Technology
(NUST)
Islamabad 44000
Pakistan
farzana.nazir@sns.nust.edu.pk

Chapter 7
Zubair Ahmad
Department of Chemistry
University of Swabi
Anbar 23561
Khyber Pakhtunkhwa
Pakistan
za3724364@gmail.com

Tahseen Arshad
Department of Chemistry
School of Natural Sciences
National University of Science and Technology
(NUST)
Islamabad 44000
Pakistan

Hassan Zeb
Department of Statistics
Islamia College University
Peshawar 25120
Khyber Pakhtunkhwa
Pakistan

Shahid Ali Khan
Department of Chemistry
School of Natural Sciences
National University of Science and Technology
(NUST)
Islamabad 44000
Pakistan
and
Department of Urology
Key Laboratory of Biological Targeting Diagnosis
Therapy and Rehabilitation of Guangdong
Higher Education Institutes
The Fifth Affiliated Hospital of Guangzhou
Medical University
Guangzhou Medical University
Guangzhou 510700
China
shahid.ali@sns.nust.edu.pk

Chapter 8
Shahid Ali Khan
Department of Chemistry
School of Natural Sciences
National University of Science and Technology
(NUST)
Islamabad 44000
Pakistan
and
Department of Urology
Key Laboratory of Biological Targeting Diagnosis
Therapy and Rehabilitation of Guangdong
Higher Education Institutes
The Fifth Affiliated Hospital of Guangzhou
Medical University
Guangzhou Medical University
Guangzhou 510700
China
shahid.ali@sns.nust.edu.pk

Zubair Ahmad
Department of Chemistry
University of Swabi
Anbar 23561
Pakistan

Zhiduan Cai
Department of Urology
Key Laboratory of Biological Targeting Diagnosis
Therapy and Rehabilitation of Guangdong
Higher Education Institutes
The Fifth Affiliated Hospital of Guangzhou
Medical University
Guangzhou Medical University
Guangzhou 510700
China

Guibin Xu
Department of Urology
Key Laboratory of Biological Targeting Diagnosis
Therapy and Rehabilitation of Guangdong
Higher Education Institutes
The Fifth Affiliated Hospital of Guangzhou
Medical University
Guangzhou Medical University
Guangzhou 510700
China

Humera Bibi, Waseeq Ur Rehman, Shah Hussain, Kausar Shaheen,
Zarbad Shah, Muhammad Asim, Shahid Ali Khan*

Chapter 1
Miscellaneous applications of hydrogels

Abstract: Hydrogels are hydrophilic polymers that retain large amounts of water in
their three-dimensional polymeric structures. Ordinary hydrogels were progressively
exchanged by artificial types due to their huge water absorption ability. The discovery
of hydrogel plays a significant role in the biomedical field without any hazard. Hydrogel has many applications in the biomedical field such as bone tissue engineering,
drug delivery, insulin delivery, nanomedicine, personal care products, contact lenses,
food packaging, and agriculture setup. Although the use of hydrogels is exceedingly
high, this chapter covers a few miscellaneous applications of hydrogels in biomedi-
cals, foods, agriculture, and catalysis.

Keywords: Hydrogel synthesis, biomedical applications, food packaging, agriculture
sector

1.1 Introduction

Hydrophilic gels are generally complexes of polymer chains that are occasionally ini-
tiated as colloidal gels in which water is used as a medium [1]. Hydrogel is a polymeric
network that shows the aptitude to expand and holds water in its structures, but it is
not soluble in water. In the past 50 years, hydrogels played a pivotal role in diverse

*Corresponding author: Shahid Ali Khan, Department of Chemistry, School of Natural Sciences, Na-
tional University of Science and Technology (NUST), Islamabad 44000, Pakistan; Department of Urology,
Key Laboratory of Biological Targeting Diagnosis, Therapy and Rehabilitation of Guangdong Higher Ed-
ucation Institutes, The Fifth Affiliated Hospital of Guangzhou Medical University, Guangzhou Medical
University, Guangzhou 510700, China, e-mail: shahid.ali@sns.nust.edu.pk
Humera Bibi, Shah Hussain, Department of Chemistry, Goverment Postgraduate College, Nowshera,
Khyber Pakhtunkhwa 24100, Pakistan
Waseeq Ur Rehman, Muhammad Asim, Department of Chemistry, Goverment Postgraduate College,
Nowshera, Khyber Pakhtunkhwa 24100, Pakistan; Department of Chemistry, Bacha Khan University,
Charsadda, Khyber Pakhtunkhwa 24420, Pakistan
Zarbad Shah, Department of Chemistry, Bacha Khan University, Charsadda, Khyber Pakhtunkhwa
24420, Pakistan
Kausar Shaheen, Department of Physics, University of Peshawar, Peshawar, Khyber Pakhtunkhwa
25120, Pakistan

https://doi.org/10.1515/9783111334080-001

technological applications [2–4]. The capability of hydrogels to hold water can be detected from the hydrophilic functional part attached to the large molecular chain, and their polymeric chain is cross-linked by cross-linker molecules. Natural hydrogels were replaced by synthetic hydrogels, owing to their high water absorption ability. Synthetic hydrogels are made from various synthetic monomers that are resistant to external stimuli [5].

Hydrogels are synthesized in a number of single- and multi-step processes, such as cross-linking of multifunctional monomers and polymerization or synthesis of molecular precursors followed by cross-linking and polymerization. Chemists develop molecular-level polymer systems that regulate a structure that has several properties including cross-linking density, bioremediation, and higher stability of chemical as well as biotic reaction to stimulants [6].

1.2 Applications of hydrogel

Hydrogels are used in various fields such as pharmaceutics (drug delivery, bone tissue engineering, wound dressing, etc.), daily life cosmetics (baby diapers, feminine napkins), and laboratory as catalysts. Some applications of hydrogels are graphically presented in Figure 1.1.

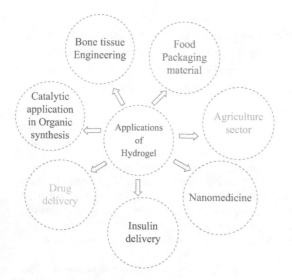

Figure 1.1: Some applications of hydrogels.

1.2.1 Hydrogels in bone tissue engineering

Since its discovery, hydrogels have been extensively researched in the realm of science, particularly in the exploration of bone tissue engineering [7, 8]. In 1960, Wichterle and Lim [9] presented a cross-linked polymer composed of 2-hydroxyethyl methacrylate monomers, and after that it got immense interest in diverse fields. It finds multiple applications in the biomedical field, for instance, bone tissue engineering [1].

Hydrogels are one of the fast emerging biomaterials for the bone tissue regeneration. Hydrogels serve as a structural carrier for the damaged bone area and enable the internal healing processes to rebuild the lost bone tissue [9]. The hydrophilic polymers that make up hydrogel are loose cross-linked systems that retain more water than their dehydrated mass. These water-retaining characteristics make hydrogels an absorbent body that allows oxygen and nutrients to diffuse swiftly inside the scaffolds [9]. Oxygen, water, pH, osmotic pressure, vitamins, and minerals are the prerequisite for the proper cell functioning, which must be fulfilled by hydrogel in bone tissue engineering. These physiological possessions make a favorable microenvironment for cells in hydrogels, which are regarded as high-quality materials for tissue repairing [11]. To enable cell infiltration and complete nutrient transport to cells, scaffolds need to have great biocompatibility, biodegradability, mechanical characteristics, and pore structure with higher interconnectivity of porosity [12, 13]. Gel scaffolds and properties of natural bones are similar, which recover the osteogenic conduct of stalk cells, and offer a good scaffolding to help in the diversity of neural stalk cells into neuron to recover the spinal cord [14, 15]. Once the cartilage is damaged, it should be surgically replaced [16] because cartilage is devoid of blood vessels, nascent cell, and lymphatic systems, so it does not repair itself. The main reason of cartilage tissue engineering is to formulate functional and scar-free tissues. The existing autologous or allogeneic bone grafts are normally useful for the cure of bone deficiencies [17–19]. Figure 1.2 indicates the role of hydrogel in bone regeneration.

The bone implants are generally reserved from the iliac crest of the patients and transplanted to the imperfect part. Autografts are energetic tissues whose cells helped in the curing process. Remodeling of autografts naturally arises by the way of osteoclast-mediated bone resorption and osteoblast-mediated bone formation. Besides this, another medication method is the use of allograft bones that is obtained from one patient and implanted into another. Allograft bones are without blood vessels, unviable tissues, an increased threat of disease transmission [17, 21]. These issues are solved by tissue engineering, for instance, nanogels combine with the elements of porcelain materials, like bone phosphate of lime, or tricalcium phosphate $Ca_3(PO_4)_2$, pentacalciumhydroxide triphosphate [$Ca_{10}(PO_4)_6(OH)_2$], demineralize bone matrix, and limestone to increase the mechanical properties [7, 10]. The research study reveals that hydroxyapatite increases the activity and feasibility of cells cultured on composites [10]. Existing cures for bones recovery positively cure bones cleft needs metallic scaffolds to aid exact bones medication. Usually, surgical-grade metal-like stainless steel

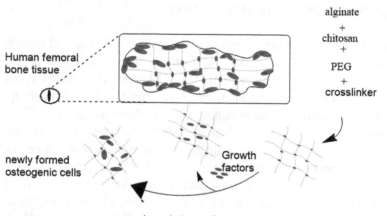

Figure 1.2: Hydrogel-assisted bone regeneration [20].

or titanium retains exceptional mechanical potency to cure defects in heavier bone as well as thigh bone, shinbone, and backbone [19]. Bioactive and osteoinductive biosensor like calcium phosphate, a major constituent of bones tissue, accelerates bone healing.

However, natural and synthetic biomaterials are useful for making injectable hydrogel frame for bone tissue engineering. Natural biomaterials used for infused hydrogel preparation include β-sitosterol, connective tissues, alginic acid, β-amyloid, keratin proteins, unfractionated heparin, chondroitin polysulfate, and hyaluronate [12, 18, 21–25]. Currently, injectable hydrogel received immense interest in the biomedical field.

1.2.1.1 Injectable hydrogel used in bone tissue engineering

1.2.1.1.1 Enzymatic hydrogel injection

Hydrogel injection is usually synthesized by an enzymatic cross-linking method that has wide characteristics such as rapid regelation, site accuracy, and proper functioning at standard biological circumstances [26–31]. Many enzymatic cross-linking injectable hydrogels used in bone tissue regeneration are peroxidase, β-lactamase, plasma amine oxidase, and horseradish peroxidase (HRP). HRP is one of the recently injectable hydrogels based on enzyme. HRP is the single-chain beta-form hemoprotein which catalyzes the carbolic acid and aminobenzene derivative in the presence of peroxide. HRP is the valuable injectable hydrogel used for maintaining the physical integrity of the wound tissues [32]. Figure 1.3 shows the synthesis of injectable hydrogels through HRP enzyme cross-linking.

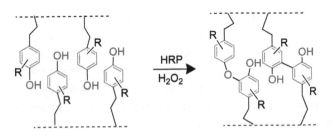

Figure 1.3: Pictorial sketch of injectable hydrogels made by the enzymatic cross-linking technique with horseradish peroxidase and H_2O_2 [33].

1.2.1.1.2 Schiff's base injectable hydrogel

Schiff bases discovered by Hugo Schiff have wide application in different field [34]. Recently, Schiff bases are reported in the synthesis of injectable hydrogels which are applied in bone tissue and cartilage regeneration [33]. Hydrogel injection based on Schiff bases has high reaction rate and ability to form imine bond [35–41]. Chitosan is used to prepare the hydrogel injection by means of Schiff-base cross-linking method to repair the cartilage tissues. Ma et al. [42] synthesized liposome-based hydrogel injection which is easily decomposed by using aldehyde/ketone-modified xanthan's gum and in-fused polymer surfactant hydrogels by using the bacterium *Xanthomonas campestris*. Xanthan gum-based surfactant hydrogels have several benefits, for example, rapidly synthesized at normal temperature, extraordinary self-curing proficiency, and ability to keep cell feasible. Scheme 1.1 shows the Schiff-base cross-linking injectable hydrogel synthesis.

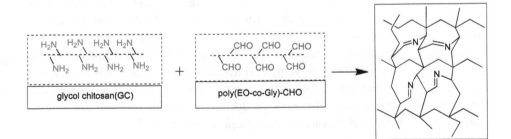

Scheme 1.1: Injectable hydrogels made by Schiff-base cross-linking between solutions of GC and poly(EO-*co*-Gly)-CHO in aqueous media [43].

1.2.1.1.3 Injectable hydrogel synthesizes by Michael addition

The attachment of a negative charge carrying carbon or a nucleophile to enones or enals compound is called Michael addition reaction. Michael addition reactions are used to synthesize the injectable hydrogel which have governable reactions time

[44–52]. Hyaluronate, β-sitosterol, and polyethylene glycol are commonly used as biosensor to synthesize hydrogel injection through the Micheal addition reactions for tendon regeneration [53–56]. Calogero and coworkers [57] synthesized two types of hyaluronate-based hydrogel injection, for instance, amino derivatives of hyaluronate, α-keratin transplant, or α/β-poly(N-2-hydroxyethyl)-D,L-aspartamide (PHEA) derived from divinyl sulfone to cure osteochondral defect. Scheme 1.2 depicts the injectable hydrogel synthesis via Michael addition reaction.

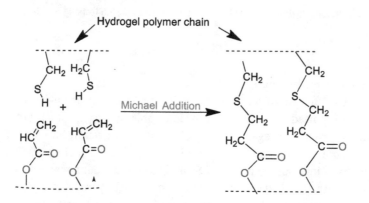

Scheme 1.2: Synthesis of injectable hydrogel by Michael addition.

1.2.1.1.4 Synthesis of injectable hydrogel based on photo-cross-linking method
Photo-cross-linking is a complicated procedure comprising three steps, such as initiation, propagation, and termination, which is activated by UV radiation [58, 59]. Papadopoulos and coworkers [58] developed a cross-linked polymeric material for the synthesis of hydrogel injection which is useful in cartilage tissue-engineering field. Pig auricular chondrocytes were condensed into polyethylene(glycol) dimethacrylate homopolymer hydrogel that consists of degradable polyethylene(glycol)-4,5-LA-dimethacrylate and nondegradable polyethylene(glycol) dimethacrylate macromere in a 50:30 M ratio. The biological and collective structures of the neocartilage specify that the feasibility, propagation, and normal production of mucopolysaccharides and hydroxyl-proline contents of the cartilage-based cell are maintained.

1.2.2 Hydrogel in drug delivery

The choice of drug delivery process is usually based on the development of the general efficiency and patient agreement. Drug delivery via topical way proposed various advantages over implantable hydrogels which are more aggressive and painful technique, vis-à-vis high cost [60, 61]. Injectable hydrogels are less painful, have a quicker retrieval period, have low prices, and have no side effects and are considered an efficient way to treat a disease [62]. Drug release through relevant path has many bene-fits and original approach. These benefits include enhancing patient obedience be-cause skin is the main protective layer of every living species and acts as a natural barrier which becomes problematic for many doses to penetrate [63]. However, scaf-folds can be inserted, and they have many advantages and keep away from the side effect caused by high medicine and quickly controlled the release of energetic com-pounds. The release of drug in response of external stimuli is shown in Scheme 1.3.

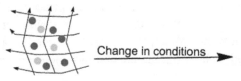

Change in conditions

Drug loaded hydrogel polyme
via adsorptionr

release of drug
from hydrogel inside
targeted organs

Scheme 1.3: Drug delivery hydrogel response to physical and chemical stimuli.

1.2.2.1 Stimuli-sensitive drug delivery hydrogels

Drug delivery hydrogels are stimuli-responsive. These include physiochemical stimuli such as physical stimuli hydrogels that are noncovalently cross-linked, and in aqueous environment they expand. The types of physically responsive hydrogels are tempera-ture, electric field, magnetic field, light, and pressure. Few examples of each type of stimuli-responsive hydrogels are presented below.

Temperature-responsive drug delivery hydrogels formed from N-trimethyl chito-san chloride polymer cross-linked with polyether and glycerol-3-phosphate [74], poly (N-vinylcaprolactam) grafted with poly(ethylene glycol) [64], poly(N-isopropylacrylamide) and aminated alginic acid [65], poly(N-vinylcaprolactam) [66], and methoxypoly(ethylene glycol)–poly(pyrrolidone-co-lactides) [67] are some examples. Similarly, pressure-respon-sive drug delivery hydrogel is formed from water-soluble synthetic linear polymers such as polyacrylamide and poly(acrylamide-hydroxyethyl methacrylate) [68]. Magnetic field-responsive drug delivery hydrogel is formed from heteropolymer hemicellulose cross-linked with O-acetyl-galactoglucomannan comprising mannose, glucose, and galactose [69], and gelatin hydrogels laden with magnetic nanoparticles [70]. Electric field-respon-sive drug delivery hydrogel is formed from five-membered heterocyclic polypyrrole poly-meric nanoparticles laden with polylactic-co-glycolic acid and polyether hydrogel [71]. Light-responsive drug delivery is formed from hydroxypropyl methylcellulose and carbo-pol hydrogels containing diclofenac-sodium chitosan microspheres [72].

Chemical stimuli hydrogels are covalently cross-linked, and the reversible or irre-versible changes and cross-linking strength in hydrogel are influenced by chemical stimuli. The type of chemical-responsive hydrogels includes pH-responsive, ionic strength-responsive, solvent composition, and molecules. For instance, pH-responsive drug delivery hydrogel is formed from carboxylated agar/tannic acid cross-linked with zinc ions [73], poly(acrylamide-co-acrylic acid) hydrogels, and so on [74]. The molecule-responsive drug delivery hydrogel is formed from N-isopropylacrylamide cross-linked with N,N-methylenebis(acrylamide) [75] and acrylamide cross-linked with polyethylene glycol [76]. Another type of stimuli-responsive hydrogel is solvent-responsive drug delivery hydrogel formed from fluorenylmethoxycarbonyl diphenylalanine [77], poly(N-isopropylacrylamide) and poly(N,N-dimethylacrylamide) mixtures [78], and poly(N-isopropylacrylamide) [79, 80]. The ionic-responsive drug delivery hydrogel is formed from 2-acrylamido-2-methylpropanesulfonic acid cross-linked with N,N-methylene(bis)acrylamide [81] and poly(N-isopropylacrylamide) cross-linked with imidazolium-based dicationic ionic liquid [82].

1.2.3 Hydrogels in nanomedicines

Nanomedicine is the rapidly growing scientific domain that utilizes the knowledge of nanotechnology to treat and cure the disease. The field of nanomedicine extends from ocular devices, biosensors, efficient medication delivery methods, and image investigation [83, 84]. The important term "nano" exactly referred to nanofillers and nanoparticles, with particle size ranging from 1 to 100 nm. Nanomedicine plays an important role and changes the purpose of nanomaterials with synergism to biological molecules. The mixing of nanomaterials and biological molecules discovered investigative apparatus and new physiological treatments [84]. In different environments, the reaction of hydrogels mainly depends on functional group of polymerize chains [85]. The use of hydrogels in nanomedicine exhibited several important characteristics such as antibacterial potential, biocompatibility, and biodegradability. These characteristics make hydrogels one of the most diverse biomaterials in nanomedicine. One example is gelatine polymer that is extracted from bones, fins, sea urchin, and so on, but mostly from porcine. Gelatine in combination with other polymers, such as chitosan, polyvinyl alcohol, and alginate, can be used to make hydrogels for various therapeutic applications [86].

1.2.4 Hydrogel used in agriculture sector

The agriculture sector required enough water resources. There are several reasons for the scarcity of water in agriculture sector such as lack of rain, improper drainage, and dumping of the waste in the water channels. Furthermore, inaccessibility of sufficient water for crops, loss of manure, and increased nutrients requirement needs appropriate supervision of water resource to sustain humidity and to raise hold water ability of the soils. The advancement of super-permeable hydrogel showed a considerable result in those areas that have insufficient water, for instance, South Asia [87–89]. Hydrogels act like a water reservoir around the roots of the plants. The swelling ability of hydrogels is up to 300–700 times to its original shape; this water holding ability makes it the most exciting materials to improve soil penetrability and enhance plant production [90–92]. Hydrogel-based large molecules are physically and chemically combined to form a complex structure. The existence of polar part in large molecules such as carboxyl, hydroxide, and carboxamide is liable to enhanced absorption aspect of the hydrogels. However, it could be divided into two parts, hydrogels based on synthetic polymer and natural polymer, where natural hydrogels own proper biocompatibility and biodegradability as compared to synthetic hydrogel. Natural hydrogels are prepared from polysaccharides and proteins such as beta-sitostrol, hyaluronate, amylum, gelatin, and connective tissue [93, 94].

1.2.5 Hydrogel in food packaging material

Hydrogel based on natural source is recyclable, bioabsorbable, and decomposable and can be manufactured on large scale, owing to organic matter abundance and current stimulating structural characteristic. The physical interactions and evading the use of toxic reagents, particularly in the cross-linking procedure, are some important parameters that should be encountered and considered before their use on large scale [95, 96]. Several foodstuff packaging materials are available in market for humidity control and food protection. Hydrogel plays a crucial role in foodstuff packaging materials and is synthesized from different compounds. For instance, poly(N,N-dimethyl acrylamide-*co*-methacryloyl sulfadimethoxine) combined with methacryloyl sulfadimethoxine monomer hydrogel is used as foodstuff cleanness indicator, to detect pH change, chemical degradation, and bacterial growth [97]. Similarly, hydrogel prepared from flavored nanoemulsions incorporated into low methoxyl pectin, whey protein isolate, orange oil, and medium-chain triglyceride oil is used as food stability and preservation of substantive materials. Hydrogel prepared from polysaccharide (starch and xanthan gum) in combination with β-carotenoid suspension enhances the bioaccessibility of lipotropic oleophilic compounds. The assimilation of lipotropic oleophilic bioactive compounds (e.g., tetraterpenoids) in foodstuff matrix improves their bioaccessibility [98].

1.2.6 Catalytic applications of hydrogels

Homogeneous catalyst in which mostly organic solvents are used are usually difficult to separate after the reaction medium as well as pose toxicity to the living organism and environment. To overcome this issue, several heterocatalytic structures and mass transport have been reported because catalytic activity is not satisfactory due to high temperature [99]. An ideal catalyst system must be harmless, inexpensive, easily available, and provide suitable reaction medium and recyclable [100]. The above-mentioned catalytic behaviors are observed in hydrogels.

The study of organic catalysts plays a vital role to develop a novel type of polymer-based biodegradable catalytic structure from hydrogels [101]. Additionally, the stimulus-responsible properties and easy recoverability enable the heterogenous catalyst separation from the reaction mixtures. The stimulus-responsive catalytic structures have a number of distinguishing characteristics, including the ability to respond to changes in temperature and pH, to changes in chemical composition and mechanical force, to changes in ionic strength, and can be conveniently examined using their nanoreactors and processes. Hydrogels are excellent substances in which active metal nanoparticles are developed/adsorbed to be used as catalyst. The empty spaces among the network of gel act as nanoreactors and give outstanding nucleation and progressive environment to the nanocomposites without any accumulation [102–104]. It is also found that gels do not undergo dissolution/solvation with the reaction medium which creates its recyclability. The metal nanoparticles develop on gels and make them as metallurgical nanocomposite catalysts.

Zhang's group [105] synthesized a thermo-responsible hydrogel-gold (16 nm size) nanoparticle stimulant from a biodegradable hydrogel-based cross-linked polymer (glyceryl methacrylate-*co-N*-isopropylacrylamide) as a chemical reactor. The hydrogel-gold nanocomposite is used for the reduction of 4-hydroxynitrobenzene, and its reduction was studied at different temperatures. The same type of catalyst is useful for catalytic reduction of *p*-nitrophenol at 25 °C, 200 s, and conversion = 65%. A general representation of 4-nitrophenol conversion to 4-amnophenol is depicted in Scheme 1.4.

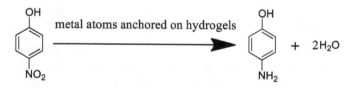

Scheme 1.4: Reduction of 4-hydroxynitrobenzen to 4-hydroxyaminobenzene.

1.3 Conclusion

Hydrogels received great attention in the biomedical application, owing to their extensive significance, great biocompatibility, and biodegradability. The enhanced strength of hydrogels tolerates a variety of applications including hydrogel used in bone tissue engineering, drug delivery, nanomedicines, food packaging, agriculture sectors, and catalysis. In this chapter we have discussed in detail the role of hydrogels in the bone tissue engineering; for instance, it provides an acceptable tissue environment, serves as a structural carrier for the damaged area, and enables the internal healing processes to rebuild the lost bone mass. In drug delivery system, hydrogels deliver the drugs or bioactive component to the specific area which might be suffered from skin penetration; therefore, an alternative tool is the injectable hydrogels which are safer. Several hydrogels are used for foodstuff packing to control humidity and keep food from ecological hazards. Hydrogels are also used in agriculture sector and can retain and release water in the drought condition. Besides, hydrogels find numerous applications in the organic reactions by stabilizing and immobilizing the nanoparticles. Hydrogels are one of the most studied materials from devices to biomedical applications. It is worth to say that no other materials have more applications than hydrogels.

References

[1]   Ahmed, E. M., et al. (2013). An innovative method for preparation of nanometal hydroxide superabsorbent hydrogel, 91(2), 693–698.
[2]   Brannon-Peppas, L. & Harland, R. S. (2012). Absorbent Polymer Technology. Elsevier.
[3]   Buchholz, F. L., Graham, A. T. J. J. W., & Sons, I. (1998). 605 Third Ave, New York, NY 10016, USA. 279, Modern superabsorbent polymer technology.
[4]   Li, Y., et al. (2013). Magnetic hydrogels and their potential biomedical applications, 23(6), 660–672.
[5]   Ahmad, Z., et al. (2022). Versatility of hydrogels: From synthetic strategies, classification, and properties to biomedical applications, 8(3), 167.
[6]   Burkert, S., et al. (2007). Cross-linking of poly (N-vinyl pyrrolidone) films by electron beam irradiation, 76(8–9), 1324–1328.
[7]   Giannoudis, P. V., Dinopoulos, H., & Tsiridis, E. J. I. (2005). Bone substitutes: An update, 36(3), S20–S27.
[8]   Sen, M. & Miclau, T. J. I. (2007). Autologous iliac crest bone graft: Should it still be the gold standard for treating nonunions?, 38(1), S75–S80.
[9]   Wichterle O, Lim D. Hydrophilic g els for biological use. Nature. 1 960; 185:117–118.
[10]   Flierl, M. A., et al. (2013). Outcomes and complication rates of different bone grafting modalities in long bone fracture nonunions: A retrospective cohort study in 182 patients, 8(1), 1–10.
[11]   Wang, P., et al. (2014). Bone tissue engineering via nanostructured calcium phosphate biomaterials and stem cells, 2(1), 1–13.
[12]   Bidarra, S. J., Barrias, C. C., & Granja, P. L. J. A. B. (2014). Injectable alginate hydrogels for cell delivery in tissue engineering, 10(4), 1646–1662.

[13]   Ossipov, D. A., Piskounova, S., & Hilborn, J. N. J. M. (2008). Poly (vinyl alcohol) cross-linkers for in vivo injectable hydrogels, 41(11), 3971–3982.

[14]   Amini, A. A. & Nair, L. S. J. B. M. (2012). Injectable hydrogels for bone and cartilage repair, 7(2), 024105.

[15]   Binetti, V. R., Fussell, G. W., & Lowman, A. M. J. J. O. A. P. S. (2014). Evaluation of two chemical crosslinking methods of poly (vinyl alcohol) hydrogels for injectable nucleus pulposus replacement, 131, 19.

[16]   Jin, R., et al. (2010). Enzymatically-crosslinked injectable hydrogels based on biomimetic dextran–hyaluronic acid conjugates for cartilage tissue engineering, 31(11), 3103–3113.

[17]   Söntjens, S. H., et al. (2006). Biodendrimer-based hydrogel scaffolds for cartilage tissue repair, 7(1), 310–316.

[18]   Ren, K., et al. (2015). Injectable glycopolypeptide hydrogels as biomimetic scaffolds for cartilage tissue engineering, 51, 238–249.

[19]   Cancedda, R., et al. (2003). Tissue engineering and cell therapy of cartilage and bone, 22(1), 81–91.

[20]   Bai, X., et al. (2018). Bioactive hydrogels for bone regeneration. Bioactive Mater, 3(4), 401–417.

[21]   Buckwalter, J. A. J. J. O. O. & Therapy, S. P. (1998). Articular cartilage: Injuries and potential for healing, 28(4), 192–202.

[22]   Gong, Y., et al. (2009). An improved injectable polysaccharide hydrogel: Modified gellan gum for long-term cartilage regeneration in vitro, 19(14), 1968–1977.

[23]   Sim, H. J., Thambi, T., & Lee, D. S. J. J. O. M. C. B. (2015). Heparin-based temperature-sensitive injectable hydrogels for protein delivery, 3(45), 8892–8901.

[24]   Wang, F., et al. (2010). Injectable, rapid gelling and highly flexible hydrogel composites as growth factor and cell carriers, 6(6), 1978–1991.

[25]   Fathi, A., et al. (2014). Elastin based cell-laden injectable hydrogels with tunable gelation, mechanical and biodegradation properties, 35(21), 5425–5435.

[26]   Lee, F., Chung, J. E., & Kurisawa, M. J. S. M. (2008). An injectable enzymatically crosslinked hyaluronic acid–tyramine hydrogel system with independent tuning of mechanical strength and gelation rate, 4(4), 880–887.

[27]   Kurisawa, M., et al. (2010). Injectable enzymatically crosslinked hydrogel system with independent tuning of mechanical strength and gelation rate for drug delivery and tissue engineering, 20(26), 5371–5375.

[28]   Park, K. M., et al. (2012). In situ SVVYGLR peptide conjugation into injectable gelatin-poly (ethylene glycol)-tyramine hydrogel via enzyme-mediated reaction for enhancement of endothelial cell activity and neo-vascularization, 23(10), 2042–2050.

[29]   Kuo, K.-C., et al. (2015). Bioengineering vascularized tissue constructs using an injectable cell-laden enzymatically crosslinked collagen hydrogel derived from dermal extracellular matrix, 27, 151–166.

[30]   Jin, R., Lin, C., & Cao, A. J. P. C. (2014). Enzyme-mediated fast injectable hydrogels based on chitosan–glycolic acid/tyrosine: Preparation, characterization, and chondrocyte culture, 5(2), 391–398.

[31]   Teixeira, L. S. M., et al. (2012). Enzyme-catalyzed crosslinkable hydrogels: Emerging strategies for tissue engineering, 33(5), 1281–1290.

[32]   Teixeira, L. S. M., et al. (2012). Self-attaching and cell-attracting in-situ forming dextran-tyramine conjugates hydrogels for arthroscopic cartilage repair, 33(11), 3164–3174.

[33]   Liu, M., et al. (2017). Injectable hydrogels for cartilage and bone tissue engineering. Bone Res, 5(1), 1–20.

[34]   Qin, W., et al. (2013). Schiff bases: A short survey on an evergreen chemistry tool, 18(10), 12264–12289.

[35]   Tan, H., et al. (2009). Injectable in situ forming biodegradable chitosan–hyaluronic acid based hydrogels for cartilage tissue engineering, 30(13), 2499–2506.

[36] Zhang, Y., et al. (2011). Synthesis of multiresponsive and dynamic chitosan-based hydrogels for controlled release of bioactive molecules, 12(8), 2894–2901.

[37] Xin, Y. & Yuan, J. J. P. C. (2012). Schiff's base as a stimuli-responsive linker in polymer chemistry, 3(11), 3045–3055.

[38] Li, Z., et al. (2015). Injectable polysaccharide hybrid hydrogels as scaffolds for burn wound healing, 5(114), 94248–94256.

[39] Jia, Y. & Li, J. J. C. R. (2015). Molecular assembly of Schiff base interactions: Construction and application, 115(3), 1597–1621.

[40] Sun, J., et al. (2013). Covalently crosslinked hyaluronic acid-chitosan hydrogel containing dexamethasone as an injectable scaffold for soft tissue engineering, 129(2), 682–688.

[41] Li, L., et al. (2015). Injectable conducting interpenetrating polymer network hydrogels from gelatin-graft-polyaniline and oxidized dextran with enhanced mechanical properties, 5(112), 92490–92498.

[42] Ma, Y.-H., et al. (2016). Biodegradable and injectable polymer–liposome hydrogel: A promising cell carrier, 7(11), 2037–2044.

[43] Cao, L., et al. (2015). An injectable hydrogel formed by in situ cross-linking of glycol chitosan and multi-benzaldehyde functionalized PEG analogues for cartilage tissue engineering. J Mater Chem B, 3(7), 1268–1280.

[44] Yang, J.-A., et al. (2014). In situ-forming injectable hydrogels for regenerative medicine, 39(12), 1973–1986.

[45] Lih, E., et al. (2008). An in situ gel-forming heparin-conjugated PLGA-PEG-PLGA copolymer, 23(5), 444–457.

[46] Censi, R., et al. (2010). In situ forming hydrogels by tandem thermal gelling and Michael addition reaction between thermosensitive triblock copolymers and thiolated hyaluronan, 43(13), 5771–5778.

[47] Lin, C., et al. (2010). Thermosensitive in situ-forming dextran–pluronic hydrogels through Michael addition, 30(8), 1236–1244.

[48] Mather, B. D., et al. (2006). Michael addition reactions in macromolecular design for emerging technologies, 31(5), 487–531.

[49] Yu, Y., et al. (2011). Novel injectable biodegradable glycol chitosan-based hydrogels crosslinked by Michael-type addition reaction with oligo (acryloyl carbonate)-*b*-poly (ethylene glycol)-*b*-oligo (acryloyl carbonate) copolymers, 99(2), 316–326.

[50] Radhakrishnan, J., Krishnan, U. M., & Sethuraman, S. J. B. A. (2014). Hydrogel based injectable scaffolds for cardiac tissue regeneration, 32(2), 449–461.

[51] Sepantafar, M., et al. (2016). Stem cells and injectable hydrogels: Synergistic therapeutics in myocardial repair, 34(4), 362–379.

[52] Kim, M., et al. (2010). Heparin-based hydrogel as a matrix for encapsulation and cultivation of primary hepatocytes, 31(13), 3596–3603.

[53] Jin, R., et al. (2010). Synthesis and characterization of hyaluronic acid–poly (ethylene glycol) hydrogels via Michael addition: An injectable biomaterial for cartilage repair, 6(6), 1968–1977.

[54] Chen, C., et al. (2013). Performance optimization of injectable chitosan hydrogel by combining physical and chemical triple crosslinking structure, 101(3), 684–693.

[55] Rodell, C. B., et al. (2015). Shear-thinning supramolecular hydrogels with secondary autonomous covalent crosslinking to modulate viscoelastic properties in vivo, 25(4), 636–644.

[56] Pritchard, C. D., et al. (2011). An injectable thiol-acrylate poly (ethylene glycol) hydrogel for sustained release of methylprednisolone sodium succinate, 32(2), 587–597.

[57] Fiorica, C., et al. (2015). Injectable in situ forming hydrogels based on natural and synthetic polymers for potential application in cartilage repair, 5(25), 19715–19723.

[58] Jeon, O., et al. (2009). Photocrosslinked alginate hydrogels with tunable biodegradation rates and mechanical properties, 30(14), 2724–2734.

[59]   Ifkovits, J. L. & Burdick, J. A. J. T. E. (2007). Photopolymerizable and degradable biomaterials for tissue engineering applications, 13(10), 2369–2385.

[60]   Li, J. & Mooney, D. J. J. N. R. M. (2016). Designing hydrogels for controlled drug delivery, 1(12), 1–17.

[61]   Cam, M. E., et al. (2020). The comparision of glybenclamide and metformin-loaded bacterial cellulose/gelatin nanofibres produced by a portable electrohydrodynamic gun for diabetic wound healing, 134, 109844.

[62]   Lee, J. H. J. B. R. (2018). Injectable hydrogels delivering therapeutic agents for disease treatment and tissue engineering, 22(1), 1–14.

[63]   Silna, E., et al. (2016). Hydrogels in topical drug delivery – A review, 2, 87–93.

[64]   Vihola, H., et al. (2008). Drug release characteristics of physically cross-linked thermosensitive poly (N-vinylcaprolactam) hydrogel particles, 97(11), 4783–4793.

[65]   Tan, R., et al. (2012). Thermo-sensitive alginate-based injectable hydrogel for tissue engineering, 87(2), 1515–1521.

[66]   Sala, R. L., et al. (2017). Thermosensitive poly (N-vinylcaprolactam) injectable hydrogels for cartilage tissue engineering, 23(17–18), 935–945.

[67]   Fu, T.-S., et al. (2018). A novel biodegradable and thermosensitive poly (ester-amide) hydrogel for cartilage tissue engineering, 2018.

[68]   Baït, N., et al. (2011). Hydrogel nanocomposites as pressure-sensitive adhesives for skin-contact applications, 7(5), 2025–2032.

[69]   Zhao, W., et al. (2015). In situ synthesis of magnetic field-responsive hemicellulose hydrogels for drug delivery, 16(8), 2522–2528.

[70]   Araújo-Custódio, S., et al. (2019). Injectable and magnetic responsive hydrogels with bioinspired ordered structures, 5(3), 1392–1404.

[71]   Ge, J., et al. (2012). Drug release from electric-field-responsive nanoparticles, 6(1), 227–233.

[72]   El-Leithy, E. S., et al. (2010). Evaluation of mucoadhesive hydrogels loaded with diclofenac sodium–chitosan microspheres for rectal administration, 11, 1695–1702.

[73]   Ninan, N., et al. (2016). Antibacterial and anti-inflammatory pH-responsive tannic acid-carboxylated agarose composite hydrogels for wound healing, 8(42), 28511–28521.

[74]   Nesrinne, S. & Djamel, A. J. A. J. O. C. (2017). Synthesis, characterization and rheological behavior of pH sensitive poly (acrylamide-co-acrylic acid) hydrogels, 10(4), 539–547.

[75]   Lu, Z. R., Kopečková, P., & Kopeček, J. J. M. B. (2003). Antigen responsive hydrogels based on polymerizable antibody Fab′ fragment, 3(6), 296–300.

[76]   Thornton, P. D., Mart, R. J., & Ulijn, R. V. J. A. M. (2007). Enzyme-responsive polymer hydrogel particles for controlled release, 19(9), 1252–1256.

[77]   Raeburn, J., et al. (2015). The effect of solvent choice on the gelation and final hydrogel properties of Fmoc–diphenylalanine, 11(5), 927–935.

[78]   Pagonis, K. & Bokias, G. J. P. B. (2007). Temperature-and solvent-sensitive hydrogels based on N-isopropylacrylamide and N,N-dimethylacrylamide, 58, 289–294.

[79]   Zhang, X.-Z., Yang, -Y.-Y., & Chung, T.-S. J. L. (2002). Effect of mixed solvents on characteristics of poly (N-isopropylacrylamide) gels, 18(7), 2538–2542.

[80]   Chang, D. P., Dolbow, J. E., & Zauscher, S. J. L. (2007). Switchable friction of stimulus-responsive hydrogels, 23(1), 250–257.

[81]   Ozmen, M. M., Okay, O. J. J. O. M. S. P. A. P., & Chemistry, A. (2006). Superfast responsive ionic hydrogels: Effect of the monomer concentration, 43(8), 1215–1225.

[82]   Zhou, X., et al. (2016). Poly (N-isopropylacrylamide)-based ionic hydrogels: Synthesis, swelling properties, interfacial adsorption and release of dyes, 48(4), 431–438.

[83]   Vashist, A., et al. (2014). Recent advances in hydrogel based drug delivery systems for the human body, 2(2), 147–166.

[84]   Vashist, A., et al. (2017). Journey of hydrogels to nanogels: A decade after.

[85] Koetting, M. C., et al. (2015). Stimulus-responsive hydrogels: Theory, modern advances, and applications, 93, 1–49.

[86] Salahuddin, B., et al. (2021). Hybrid gelatin hydrogels in nanomedicine applications. ACS Appl Bio Mater, 4(4), 2886–2906.

[87] Kalhapure, A., et al. (2016). Hydrogels: A boon for increasing agricultural productivity in water-stressed environment, 1773–1779.

[88] Vundavalli, R., et al. (2015). Biodegradable nano-hydrogels in agricultural farming-alternative source for water resources, 10, 548–554.

[89] Jayaramudu, T., et al. (2019). Swelling behavior of polyacrylamide–cellulose nanocrystal hydrogels: Swelling kinetics, temperature, and pH effects, 12(13), 2080.

[90] López-Velázquez, J. C., et al. (2019). Gelatin–chitosan–PVA hydrogels and their application in agriculture, 94(11), 3495–3504.

[91] Montesano, F. F., et al. (2015). Biodegradable superabsorbent hydrogel increases water retention properties of growing media and plant growth, 4, 451–458.

[92] Milani, P., et al. (2017). Polymers and its applications in agriculture, 27, 256–266.

[93] Vasquez, J. M. G. & Tumolva, T. P. (2015). Synthesis and characterization of a self-assembling hydrogel from water-soluble cellulose derivatives and sodium hydroxide/thiourea solution. In 11th Engineering research and development for technology conference.

[94] Sannino, A., Demitri, C., & Madaghiele, M. J. M. (2009). Biodegradable cellulose-based hydrogels: Design and applications, 2(2), 353–373.

[95] Batista, R. A., et al. (2019). Hydrogel as an alternative structure for food packaging systems, 205, 106–116.

[96] Kabiri, K., et al. (2009). Residual monomer in superabsorbent polymers: Effects of the initiating system, 114(4), 2533–2540.

[97] Baek, S., et al. (2017). Preparation and characterization of pH-responsive poly (N,N-dimethyl acrylamide-co-methacryloyl sulfadimethoxine) hydrogels for application as food freshness indicators, 120, 57–65.

[98] Park, S., Mun, S., & Kim, Y.-R. J. F.-R. I. (2018). Effect of xanthan gum on lipid digestion and bioaccessibility of β-carotene-loaded rice starch-based filled hydrogels, 105, 440–445.

[99] Ikegami, S. & Hamamoto, H. J. C. R. (2009). Novel recycling system for organic synthesis via designer polymer-gel catalysts, 109(2), 583–593.

[100] Anastas, P. T., Kirchhoff, M. M., & Williamson, T. C. J. A. C. A. G. (2001). Catalysis as a foundational pillar of green chemistry, 221(1–2), 3–13.

[101] Al-Ghamdi, Y. O., et al. (2022). Polymers blended peanuts activated carbon composite hydrogels fabricated Ag NPs as dip-catalyst for industrial dyes discoloration in aqueous medium. Ind Crops Prod, 188, 115588.

[102] Wang, Z., et al. (2007). Design of polymeric stabilizers for size-controlled synthesis of monodisperse gold nanoparticles in water, 23(2), 885–895.

[103] Palioura, D., et al. (2007). Metal nanocrystals incorporated within pH-responsive microgel particles, 23(10), 5761–5768.

[104] Pong, F. Y., et al. (2006). Thermoresponsive behavior of poly (N-isopropylacrylamide) hydrogels containing gold nanostructures, 22(8), 3851–3857.

[105] Jiang, X., et al. (2007). Thermoresponsive hydrogel of poly (glycidyl methacrylate-co-N-isopropylacrylamide) as a nanoreactor of gold nanoparticles, 45(13), 2812–2819.

Laiba Maryam, Asma Gulzar, Mudassir Iqbal*

Chapter 2
An overview of the mechanical behavior of hydrogels

Abstract: This chapter is intended to explain the versatile and compelling properties of hydrogels with a deep insight into their mechanical behavior. Hydrogels being polymeric in nature are the "smart materials" having properties like swelling, inhomogeneity, biocompatibility, biodegradability, stimuli responsiveness, pH and temperature sensitivity, and especially mechanical properties. Due to these inherent properties, hydrogels find applications in many fields such as medicine, drug delivery, energy storage, water sustainability, sensing applications, and electrochemical devices. Mechanically they can be tested on different scales, both at the macroscale and microscale. Indentation testing is a multiscale method used to measure the mechanical strength of hydrogels and can be categorized as microindentation or nanoindentation. The mechanical behavior of hydrogels is extensively discussed by focusing on the properties that impart mechanical strength. These properties include swelling–deswelling behavior (which is due to the absorption of water), rubber elasticity, viscoelasticity, creep behavior, hyperelasticity, and self-healing properties to repair cracks and fractures in 3D networks. All the factors that affect the mechanical strength of hydrogels are also discussed.

Keywords: Hydrogels, mechanical properties, smart materials, mechanical testing

2.1 Introduction

The term "hydrogel" was first introduced in 1960 by Wichterle and Lím when they successfully synthesized poly(2-hydroxyethyl methacrylate) and applied it in the contact lens industry. Hydrophilic gels or hydrogels are three-dimensional (3D) networks of polymer chains that have the water-binding ability due to hydrogen bond formation. This promising hydrophilicity is because of various hydrophilic groups present as branched moieties along the polymer chain such as -OH, -COOH, $-NH_2$, -CONH, $-CONH_2$,

Acknowledgments: We acknowledge the contribution and support from the Department of Chemistry, School of Natural Sciences, National University of Science and Technology, Islamabad.

*Corresponding author: Mudassir Iqbal, Department of Chemistry, School of Natural Sciences (SNS), National University of Sciences and Technology (NUST), Islamabad 44000, Pakistan,
e-mail: mudassir.iqbal@sns.nust.edu.pk
Laiba Maryam, Asma Gulzar, Department of Chemistry, School of Natural Sciences (SNS), National University of Sciences and Technology (NUST), Islamabad 44000, Pakistan

https://doi.org/10.1515/9783111334080-002

and SO_3H. Depending on the polymer concentration, the water content of hydrogels can range from at least 10% to a maximum of 99%. Water in hydrogels can be categorized into three types: free water, semibound, and water adhered to the polymer chain [1, 2].

This water absorption defines the swelling ability of hydrogels, makes them elastic, and imparts intriguing relaxation behaviors which are the characteristics of their gel state. Swelling media, hydrophilicity of the attached moieties, and cross-linked bonding strength are the factors that govern the swelling character. Cross-linking is the prime factor that maintains the 3D network in the swollen state. Usually, mass of the polymer is less than the mass of water, depending on the network density and polymer properties. Due to their soft nature, they have low mechanical strength, but they can be functionalized in many ways to improve their mechanical properties [3].

External stimuli (either physical or chemical) can cause some drastic transitions such as sol–gel and volume phase transitions. Physical stimuli may include temperature, solvent quality and composition, magnetic and electric fields, pressure, and light intensity, while chemical stimuli include chemical compositions, ions, and pH. Their response to external stimuli is primarily regulated by charge density, degree of cross-linking, pendant chains, and the nature of the monomer. The magnitude of response is also directly correlated to the applied external stimulus.

There are numerous forms of hydrogels, including slabs, beads, cryogels, microgels, nanogels, membranes, and aerogels. Owing to their versatile, idiosyncratic multifunctional properties, hydrogels are considered "smart materials." Their peculiar properties such as elasticity, viscoelasticity, fluffiness, softness, super-absorbency, hydrophilicity, biodegradability, and biocompatibility make them suitable for various biomedical applications, their role in tissue engineering, and drug delivery. They can also be employed for environmental remediation by removing hazardous materials and heavy metals from water. This is an eco-friendly way of filtering metals from wastewater with the help of hydrogel materials from a plant source. Their stability against hydrolysis and permeability toward oxygen makes them appealing in the lens industry. For example, many silicone-based hydrogels are in commercial use for making contact lenses. In some instances, they are used as sensors such as pH sensors, pollution sensors, and radiation sensors [4].

Recently, advances in hydrogel technology enable them to have applications in water sustainability and energy storage. Liquid electrolytes that cause the device leakage are replaced with hydrogel electrolytes due to their inherent flexibility and semisolid phase. Hydrogels enhance flexibility, stretchability, and elasticity of electrochemical devices, enabling them to withstand bending, folding, twisting, and stretching. When compared to conventional electrochemical devices, these features are extremely destructive. Hydrogels are also capable of self-healing, which is important for portable and wearable electronic equipment. Devices with self-healing or damage control capabilities are perfect for smart and lightweight electronics. These fascinating applications grabbed hydrogels' striking attention nowadays [5, 6].

2.2 Classification of hydrogels

There are different aspects on which hydrogels can be classified. Depending on the source from which they originated, they may be natural, synthetic, or an amalgam of both. Based on the polymer composition, they are classified as homopolymeric hydrogel composed of only one monomer which is the structural unit, copolymeric hydrogel composed of two or more different kinds of monomers having at least one hydrophilic constituent, and the most predominant one is multipolymer interpenetrating polymeric hydrogel that contains two independently cross-linked units in the network. Hydrogels can also be categorized based on the configuration of their physical structure. They may be noncrystalline (amorphous), semicrystalline, or crystalline. Cross-linking behavior of hydrogels defines them as chemicals having permanent junctions, physical having some transient junctions between them, or hybrid. Depending on the polymerization technique employed, they may appear as film, matrix, or microsphere. Because of the electrical charge of the network, they may be neutral or nonionic, ionic (having anions and cations), polybetaines (i.e., zwitterionic), and ampholytic consisting of acidic and basic groups also known as amphoteric electrolytes. According to stimuli responsiveness, they respond to all physical, chemical, and biochemical responses. Hydrogels can also be extensively classified based on their preparation method.

2.3 Preparation of hydrogels

Various preparation techniques used in synthesizing hydrogels are bulk polymerization, suspension polymerization, solution casting, free-radical polymerization, and irradiation method.

In bulk polymerization, one-to-many types of monomers are formulated in the presence of a minute amount of a linker, that is, cross-linker, and an initiator to form hydrogels. With the help of this polymerization, hydrogels such as particles, emulsions, films, rods, and membranes can be generated.

With the cross-linking substances, neutral or ionic monomers are combined during solution polymerization. Ultraviolet rays or the redox initiator method can be used to start the polymerization thermally. To achieve the activation energy of the reaction solvent is used such as benzyl alcohol, ethanol, or water–ethanol. To get rid of the initiator, cross-linkers, unreacted monomers, oligomers, and other contaminants, the produced hydrogels must be washed with water [7].

Free radical polymerization is an alternative technique that can be used with monomers having functional groups like vinyl lactams, acrylates, or amides that undergo radical polymerization. Stages of radical polymerization include initiation, propagation, chain transfer, and termination [8].

Suspension polymerization is also called inverse suspension as it substitutes the conventional oil-in-water method to water-in-oil. In the hydrocarbon phase, a mixture of initiator and the monomers is propagated. Agitation speed, dispersant type, rotor design, and the viscosity of the monomer solution controlled the particle size and the form of a resin. This method has been proven to be very beneficial as the end product is in the form of beads (microspheres) or powder, so there is no need for grinding [9, 10].

Hydrogels of unsaturated compounds can be prepared by irradiating them with high-energy radiations like electronic beams and gamma rays resulting in the generation of radicals. The hydroxyl radicals are formed by radiolysis of H_2O molecules which then strike the polymer chains resulting in the formation of macroradicals. Then a covalent bond is formed by recombining different polymer chains and ultimately a cross-linked networked structure is obtained [11].

2.4 General properties of hydrogels

Hydrogels are very advantageous and ubiquitous due to their unique characteristic properties. The major properties of hydrogels are explained below.

2.4.1 Swelling behavior of hydrogels

The most eminent property of hydrogels which is responsible for the absorption and retention of water in the polymeric network is swelling. The Flory–Reihner theory states that swelling plays a key role in the elasticity of polymeric chains and their affinity with water molecules. Three steps are involved in the swelling of hydrogels:
(i) The diffusion of primary bound H_2O into the network of hydrogel
(ii) Relaxation of polymer chains by secondary bound H_2O
(iii) Expansion of hydrogel network by additional H_2O which is free water

The formula by which equilibrium swelling of hydrogels can be calculated is

$$\text{swelling ratio} = \frac{W_t - W_d}{W_d}$$

whereas W_t corresponds to swollen hydrogel weight and W_d refers to dried hydrogel weight after freeze-drying. Molecular weight and cross-linking are the two key features that govern the swelling of hydrogels. Cross-linking not only restricts the dissolution but also preserves the dimensions of the 3D network. The swelling of hydrogels is very sensitive to external factors like pH, temperature, and ion concentrations, which can cause the hydrogel to collapse or transition into a different phase [12, 13].

2.4.2 Inhomogeneity of hydrogels

The inhomogeneous distribution of cross-link density causes anomalous spatial gel inhomogeneity which is the nonideal characteristic of hydrogels. This undesirable feature drastically diminishes the strength and optical clarity of hydrogels. As fluctuations in the spatial concentrations are directly related to the gel inhomogeneity, various scattering methods such as small-angle neutron scattering, small-angle X-ray scattering, and light scattering have been used for the investigation of homogeneity. The same concentrations of scattering intensities from the gels and semidilute solution of the same polymer are compared to calculate the inhomogeneity. The simultaneous increase in the cross-link density and network deformities will result in increased inhomogeneity in hydrogels while it decreases with the increase in the degree of ionization due to the electrostatic repulsions, mobile counterions, and the Donnan potential. The swelling of gels may also have some degree of effect on the inhomogeneity of hydrogels [14, 15].

2.4.3 Biological properties of hydrogels

The high water-retaining capacity of hydrogels enables them to shelter cells and proteins without changing their nature and characteristic properties and thus enables them to be an excellent candidate for biomedical applications. Their unique structures, structural diversity, enhanced biocompatibility, biodegradability, high water swellability, high oxygen permeability, ease of loading and release of drugs, and nontoxicity boosted their biological properties. The biological properties of hydrogels include antioxidant, anti-viral, antibacterial, antifungal, and anti-inflammatory properties. Poly(tannic acid) hydrogels are a strong contender for the treatment of chronic wounds due to their high antioxidant capacity as they are natural polyphenols. To make hydrogels effective against viruses, nanoparticles of some metals and nonmetals such as Ag, ZnO, CeO$_2$, TiO$_2$, FeO$_x$, ZnS, and CdSe are incorporated. The antibacterial properties associated with peptides make them a promising candidate to cure bacterial infections. Some hydrogel-based antibiotics are gentamicin, ciprofloxacin, vancomycin, and so on. Amphotericin B is a compelling antifungal agent which was used to synthesize antifungal biohydrogel by its absorption into the dextran-based hydrogel. This broad spectrum of antifungal activity enables this biohydrogel to destroy fungi within 2 h of contact. Ascorbyl palmitate, a highly bioavailable, fat-soluble vitamin C derivative, was used to create hydrogel microfibers that target inflammation. Dexamethasone, an anti-inflammatory corticosteroid, was added to the manufactured biohydrogel microfibers, which were then used as a medication delivery method for the inflamed colon [16, 17].

2.4.4 Stimuli responsiveness of hydrogels

Hydrogels are responsive to environmental stimuli that may be internal (chemical or biological) or external (physical except temperature). Physical factors include light and temperature, chemical stimuli include ionic concentrations and pH, and biological stimuli include enzymatic actions. In addition to these, shape memory hydrogel is a specific stimulus-responsive smart hydrogel that has two features: (i) a permanent shape and (ii) a chemical or physical coding that enables it to be restored to its original shape [17, 18].

2.4.5 pH and temperature sensitivity of hydrogels

Covalently cross-linked acidic or basic moieties that are present in the structure respond to the pH changes of hydrogels. High pH causes the hydrogels to swell at a better rate because of the presence of anionic groups that promoted electrostatic repulsion. Instead, cationic hydrogel lacks this property. Swelling and deswelling of the temperature-sensitive hydrogel can be observed by alteration of the temperature that is because of polymer–water and polymer–polymer interaction which is temperature sensitive. Illustration of poly(N-isopropyl acrylamide) hydrogel is a temperature-sensitive hydrogel. Hydrogels can either be positively or negatively temperature sensitive. When hydrogels undergo swelling at elevated temperature than the upper critical solution temperature, they are positive temperature-sensitive hydrogels. While swelling at a temperature is lower than the lower critical solution temperature, they are negative temperature-sensitive hydrogels [18, 19].

2.4.6 Mechanical properties of hydrogels

Hydrogels' mechanical strength is crucial for applications in the pharmaceutical and biological fields. For appropriate physiological performance in a variety of contexts, including biomedical applications, ligament and tendon healing, cartilage replacement, wound dressing, drug delivery matrix, and tissue engineering, hydrogels' mechanical strength must be evaluated. It is crucially important for a hydrogel to maintain its appearance during the process of therapeutic drug delivery for a specific duration [20, 21]. The elasticity and viscoelasticity of hydrogels result in a dual mechanical behavior. These can be explained at the microscopic level in terms of how the hydrogel polymer matrix adapts to applied stress or strain. Hydrogels are very fragile materials with low mechanical strength, unlike other polymers like plastics but their mechanical strength can be improved by functionalizing them. Let us discuss the mechanical behaviors of hydrogels in detail [22, 23].

2.4.7 Mechanical testing of hydrogels

Mechanical testing of hydrogels can be done at different scales and can be categorized into three types:
1. Macroscale testing
2. Indentation testing
3. Microscopic testing

2.4.7.1 Macroscale testing

The macroscopic properties of hydrogels can be studied by using a versatile machine known as UTM (universal testing machine). UTM can carry out a broad spectrum of mechanical testing, which includes fracture tests, compression and uniaxial tension tests, stress relaxation, and cyclic loading tests. UTM typically composed of a rigid load frame to support the machine, a crosshead which is programmable to regulate movement, a load cell to detect force, and test fixtures to hold specimens of various forms [24, 25].

Tension testing or tensile tests of hydrogels can be carried out in both hydrated and dry forms, but the specimen should be in a dumbbell shape. In order to avoid the sudden change from the linear to the nonlinear domain during load-controlled tests, tension test of hydrogel is typically conducted under displacement control. Force–displacement data is recorded during tension testing which is then converted into stress–strain data.

During a compression test, a cylindrical sample is pressed by applying compressive pressure with the help of parallel plates on UTM. Stress–strain responses of the hydrogels and their compressive modulus can be determined by performing uniaxial compression tests [26].

An essential mechanical feature for hydrogels is toughness of fracture, which denotes a material's capacity to withstand enlargement under applied load or crack propagation. Since failure stress significantly depends on the size and shape of the sample, they do not provide accurate estimations but rather give a rough idea of fracture toughness. The primary goal is the measurement of the key relationship between crack size, applied load, and fracture toughness, that is, material resistance that will induce the crack growth [26, 27].

2.4.7.2 Indentation testing

Indentation is a method used for examining the mechanical characteristics of hydrogels from millimeter (m) to nanoscale (n). A rigid probe with a specific geometry is pressed into the test sample during a traditional indentation test with a predetermined load or depth, and the related load (P), indentation depth (h), and time (t) are all recorded. Empirical constitutive models are used to examine the force–depth–time (P–h–t) data in order to find many significant material properties. Indentation testing is more beneficial than macroscale testing because it requires very little preparation of sample and can use specimens of any shape, as hydrogels are tested in containers in which they are synthesized without being attached to the instrument which is one of the significant problems during tension and compression testing [28, 29]. An indentation can be either a nanoindentation or a microindentation, depending on the length scale of the indentation depth. According to the Hertz model, the corresponding indentation force–displacement relation is given as

$$P = \frac{4\sqrt{R}}{3} \frac{E_h}{1 - v^2} h^{3/2}$$

where R is the tip radius of the spherical indenter, E_h is the elastic modulus, and V is the Poisson's ratio. The elastic modulus can be determined by fitting it to the experimental data.

2.4.7.3 Microscopic testing

A nanoscale microscopic level of mechanical testing of hydrogels can be carried out by using atomic force microscopy (AFM). The only difference in AFM and indentation testing is that force is measured only by deflecting the probe into calibrated cantilever beam rather than stimulating the probe directly into the sample. In this way, the deformation of the sample can be avoided. Two most important noncontact microscopic techniques that are employed to determine the physical properties of hydrogels such as diffusivity are light scattering and fluorescence recovery after photobleaching [27].

2.4.8 Swelling–deswelling behavior

Hydrogels being hydrophilic materials are capable of absorbing an ample amount of water, resulting in the high-water content in their structures. The main cause of this swelling is the formation of hydrogen bonds between polymeric chains and water which reduces the energy. Many theories have been developed that explain the swelling behavior of hydrogels at both macroscopic and microscopic levels. According to mechanics, when a hydrogel attains the swelling equilibrium, then the external and internal chemical potential also establishes an equilibrium between them ($\mu_{in.} = \mu_{ext.}$). Chemical potential can be calculated as

$$\frac{\partial W\ (F_S,\ \mu)}{\partial\ F_S} = 0$$

where "W" is the free energy of hydrogels with respect to the swelling deformation gradient "F_S" and chemical potential "μ." This equation successfully defines chemical potential as the stretching force of hydrogels but failed to determine the chemical potential at the microscopic level. Katchalsky et al. developed the microscopic view of the mechanism of swelling. According to Katchalsky et al., the balance between the internal stress of polymer chains and the osmotic pressure on chains from the solvent environment determines the swelling equilibrium of hydrogels, that is ($P_{in.} = P_{ext.}$). Three parts constitute the total free energy of the polymeric chains: the stretch part F_{str}, the electrostatic part F_{el}, and the mixed part F_{mix}. As these parts are related to the gel volume, swelling equilibrium equation can be expressed as follows:

$$\sum i\left(\frac{\partial F^i_{str}}{\partial V} + \frac{\partial F^i_{el}}{\partial V} + \frac{\partial F^i_{mix}}{\partial V}\right) = P_{ext}$$

The swelling ratio and the mechanical behavior can be anticipated using these swelling theories [30, 31].

2.4.9 Rubber elasticity

Rubbers are polymers that deform almost instantly upon stress, but their deformation is completely reversible. When hydrogels behave like rubber materials, their mechanical behavior mainly depends on the structure of polymer network. The state of rubber elasticity of hydrogels can be explained in both swollen and unswollen states. In unswollen state, the equation of state for rubber elasticity can be derived by combining classical thermodynamics and statistical analysis and is given as

$$f = \frac{\partial U}{\partial L_{TV}} + T\frac{\partial F}{\partial T_{LV}}$$

where f is the retractive force in response to a tensile force, U is the internal energy, "L" is the length, "V" is the volume, and "T" is the temperature. Figure 2.1 represents the same:

By assuming "N" number of network chains in hydrogel medium, ΔF (total change) can be evaluated as follows:

$$\Delta F - \int f dr \frac{3NkT}{<r_f^2>} \int\limits_{\sqrt{<r_f^2>}}^{\sqrt{<r^2>}} r dr$$

The shear modulus of hydrogel network in the unswollen state can be obtained using constitutive relationships;

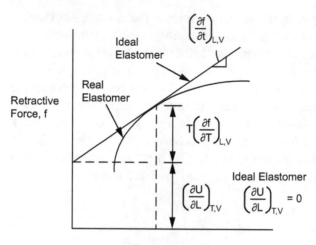

Figure 2.1: The thermodynamic analysis of equation of state for rubber elasticity. Reproduced after permission [22].

$$G = \frac{\rho RT}{M_C} \frac{<r_0^2>}{<r_f^2>} \left(1 - 2\frac{\overline{M_C}}{\overline{M_N}}\right)$$

And stress is given as

$$\tau = G\left(\lambda - \frac{1}{\lambda^2}\right)$$

In the swollen state, for a perfect gel network, the relationship between the degree of swelling and shear modulus is depicted as

$$G_{sw} = G_0 Q^{\frac{-1}{3}}$$

where $Q = V\,V_0$ is the ratio between the volume in the unswollen state and the swollen state, and G is the shear modulus in the unswollen state. It can be shown that even in the swollen state, the shear stress still has the same relationship to the strain:

$$\sigma = G_{sw}\left(\lambda - \frac{1}{\lambda^2}\right)$$

Figure 2.2: Rheometer.

2.4.10 Viscoelasticity

The ability of a substance to exhibit both elastic and viscous behavior when deformed is known as viscoelasticity. This means they can be stretched just like the elastic materials but cannot revert to their original shape after the removal of tensile force. Viscoelastic theory of hydrogels deals with elasticity, molecular motion, and flow in the polymeric network. The nature of applied mechanical stress is directly proportional to the magnitude of viscoelastic response of the hydrogel. Time-dependent mechanical behavior of hydrogels or any soft material can be determined with the help of viscoelasticity. The equation of isotropic linear viscoelastic material can be shown as

$$\sigma_{ij}(t) = \int_0^t \left[2G(t-\tau)\frac{\partial \varepsilon_{ij}(\tau)}{\partial t} + K(t-\tau)\delta_{ij}\frac{\varepsilon_{kk}(\tau)}{\partial t} \right] d\tau$$

where σ and ε represent stress and strain, respectively, $G(t)$ is the shear relaxation modulus, and $K(t)$ is the bulk relaxation modulus. Correspondingly, the strain tensor can be expressed as

$$\varepsilon_{ij}(t) = \int_0^t \left[\frac{1}{2}J\ (t-\tau)\frac{\partial \sigma_{ij}(\tau)}{\partial t} + \frac{1}{3}M\ (t-\tau)\delta_{ij}\frac{\partial \varepsilon_{kk}(\tau)}{\partial t} \right] d\tau$$

where $J(t)$ is the shear creep compliance and $M(t)$ is the volume creep compliance.

The physical models that are employed to explain the viscoelasticity of hydrogels include Maxwell models and Kelvin–Voigt model. The viscoelasticity can be calculated with the help of rheometer which measures the flow of matter and extent of deformation under stress or strain (Figure 2.2).

2.4.11 Creep behavior

The strain on a viscoelastic material is time-dependent under constant stress. The creep compliance is known to be the ratio of time-dependent strain to the applied shear mechanical stress. Kelvin–Voigt model is used to describe this creep behavior of hydrogels. The creep strain is given as

$$\varepsilon(t) = \sigma C_0 + \sigma C \int_0^\infty f(\tau)\left(1 - e^{\frac{t}{\tau}}\right) d\tau$$

where "σ" is the applied stress, "C_0" is the instantaneous creep compliance, "C" is the creep compliance coefficient, "τ" is the relaxation time, and $f(t)$ is the distribution of relaxation times. Figure 2.3 shows the dependence of stress and strain.

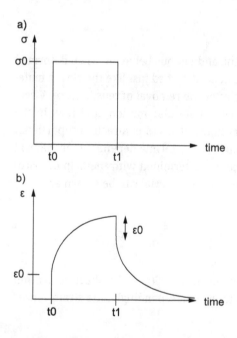

Figure 2.3: Illustration of creep behavior showing stress–strain curves in viscoelastic polymers. Reproduced after permission [3].

2.4.12 Hyperelasticity

The impact of the network randomness is required to examine the hyperelastic behavior of hydrogels from a network perspective. Various simulations are done by constructing different 2D models that may be full-atom, abstract, or realistic model and compared them with ideal network assumptions to evaluate the randomness. Many hyperelastic models have already been designed for single-mode deformation, and now recent approaches are trying to develop hyperelastic models for multiple deformation modes. The key factor that governs the applicability of these hyperelastic models is the gel concentration [32, 33].

2.4.13 Self-healing of hydrogels

One of the alluring features of mechanical behavior of hydrogels is self-healing which is the repairing of cracks in the bulk polymer network. Due to the recombination and dissociation of chemical or physical bonds, self-healing is the reversible process, and it initiates and stimulates the physical and chemical reactions in the damaged area due to the functional moieties in polymer chains.

Through hydrogen bonding which is a noncovalent molecular interactions including ionic electrostatic interactions, metal coordination, and mixed intermolecular interactions, physically self-healing hydrogels can reconstruct their 3D network structures. Noncovalent methods of self-healing are easy, simple, rapid, and reproducible [34, 35].

Utilizing covalent linkages including imine, disulfide, acylhydrazone, and boronate ester bonds, self-healing hydrogels restore the network chemically. External factors such as temperature, current, and pH are essential to initiate the chemical self-healing of hydrogels [36].

2.5 Factors affecting the mechanical strength of hydrogels

2.5.1 Effect of solvent water

The effect of water on polymer network can be extensively studied because polymer–water interactions are one of the prime factors that governs the mechanical behavior of hydrogels. High-water content is sustained in hydrogels due to hydrogen bonding. Water, apart from providing extra enthalpy and entropy, dramatically increases the viscosity of hydrogels which in turn provides exceptional resistance to deformation. Water layers only experience shear deformation when bulk hydrogels are being deformed.

To achieve the shear deformation of the water layer by shifting the top boundary in models, two shearwater models with various water layer thicknesses are built. Water molecules produce a distinct layered pattern after relaxing, with a thickness of 0.62 nm, or little over twice the diameter of a water molecule. The abrupt increase in Young's modulus of hydrogels with decreasing water content is studied by considering the deformation of bulk hydrogels at microscopic level. It is clearly demonstrated that water in hydrogels, especially those with low water content, gives additional resistance to deformation.

2.5.2 Effect of cross-linking density

The mechanical strength of the hydrogels strongly depends on the cross-links in polymer network. Increasing the cross-linking density will increase the strength of the hydrogel. By increasing the amount of cross-linking agent, cross-linking density and glass-transition temperature can be increased which in the swollen and rubbery state increase the mechanical strength of the polymer network by elevating the values of shear modulus.

2.5.3 Effect of monomer concentration

During the preparation of hydrogels, changing the composition of comonomer can increase the relative mechanical strength of the final hydrogel formed. The stiffness of the polymeric backbone can be boosted by functionalizing the polymer with different moieties that impart more hydrophilicity to polymers such as methacrylates or acrylates which can be replaced in the backbone. It also depends on the type of monomers, that is, cationic or anionic. Weak hydrogels are formed when all the constituting monomers are either anionic or cationic. This means the maximum degree of swelling occurs when all the ionic monomers are either positively charged or negatively charged.

2.5.4 Effect of degree of swelling

Altering the monomer concentration, the amount of cross-linking agent, and reaction conditions diminishes the degree of swelling, and consequently this will improve the mechanical Properties of hydrogel. As the polymer swells, the weakening of polymer occurred due to reduction in the glass transition temperature. Mechanical properties can be significantly affected by manipulating the amount of swelling by adjusting external factors such as temperature, ionic strength, pH, or nature of swelling agent.

2.5.5 Effect of polymerization conditions

Conditions used for reactions such as temperature, reaction type, and type or amount of solvent also suggest the properties of the final product formed during the polymerization. The prime factor is the amount of solvent and cross-linking agents. If the amount of solvent is large, then cycles are formed instead of cross-links which decrease the density of cross-linking, which in turn reduces the mechanical property of the hydrogel. In photopolymerization, light intensity also affects the mechanical strength. For example, the number of increased double bond conversion can reduce solubility and increases the cross-linking density and hence mechanical strength is improved.

2.6 Conclusion

This chapter is the collection of basic concepts of hydrogels explaining their synthetic approach, and their classification based on the linking and bonding nature with the detailed explanation of properties, especially mechanical properties, and the factors affecting its 3D nature with hydrogels' modeling. Due to the immense use and applications of hydrogel basis on the mechanical property, which is very sensitive and depends on factors like solvents, concentrations, and conditions, it is important to keep these factors precisely and as per the requirements to use it as a drug delivery and the tissue engineering as it decides the fate of the cell which is the unit of life. Capillary alleviation of hydrogels is very sensitive to the solvent as it could be the main reason for the shrinkage of 3D structure and play a crucial role in cell's degradation, deformation, and cartilage extracellular matrix concentration. Thus, the study of hydrogels' nature and model illustrations helps in controlling the properties and applications of in vivo modeling of cells. Further studies and research in the field of ionically conductive hydrogels can prove to be a success in conductive and electronic sensing devices and cell imaging with hybrid hydrogels that can be used to make artificial soft skins and robotic bodies.

References

[1] Aghahosseini, H., Ramazani, A., Azimzadeh Asiabi, P., Gouranlou, F., Hosseini, F., Rezaei, A., . . . & Woo Joo, S. (2016). Glucose-based biofuel cells: nanotechnology as a vital science in biofuel cells performance. Nanochemistry Research, 1(2), 183–204.

[2] Ahmed, E. M. (2015). Hydrogel: Preparation, characterization, and applications: A review. J Adv Res, 6(2), 105–121.

[3] Boardman, P. Modelling the Mechanical Properties of Hydrogel.

[4] Beckett, L. E., Lewis, J. T., Tonge, T. K., & Korley, L. T. (2020). Enhancement of the mechanical properties of hydrogels with continuous fibrous reinforcement. ACS Biomaterials Science & Engineering, 6(10), 5453–5473.

[5] Bahram, M., Mohseni, N., & Moghtader, M. (2016). An Introduction to Hydrogels and some Recent Applications, in Emerging Concepts in Analysis and Applications of Hydrogels. IntechOpen doi: 10.5772/64301, S.NO: 978-953-51-2509-9.

[6] Bashir, S., Hina, M., Iqbal, J., Rajpar, A. H., Mujtaba, M. A., Alghamdi, N. A., . . . & Ramesh, S. (2020). Fundamental concepts of hydrogels: Synthesis, properties, and their applications. Polymers, 12(11), 2702.

[7] Cai, M. H., Chen, X. Y., Fu, L. Q., Du, W. L., Yang, X., Mou, X. Z., & Hu, P. Y. (2021). Design and development of hybrid hydrogels for biomedical applications: Recent trends in anticancer drug delivery and tissue engineering. Frontiers in Bioengineering and Biotechnology, 9, 630943.

[8] Caliari, S. R., & Burdick, J. A. (2016). A practical guide to hydrogels for cell culture. Nat Methods, 13(5), 405–414.

[9] Castilho, M., Hochleitner, G., Wilson, W., Van Rietbergen, B., Dalton, P. D., Groll, J., . . . & Ito, K. (2018). Mechanical behavior of a soft hydrogel reinforced with three-dimensional printed microfibre scaffolds. Scientific reports, 8(1), 1245.

[10] Chamkouri, H., & Chamkouri, M. (2021). A review of hydrogels, their properties and applications in medicine. Am J Biomed Sci Res, 11(6), 485–493.

[11] Carvalho, I. C., Medeiros Borsagli, F. G., Mansur, A. A., Caldeira, C. L., Haas, D. J., Lage, A. P., . . . & Mansur, H. S. (2021). 3D sponges of chemically functionalized chitosan for potential environmental pollution remediation: biosorbents for anionic dye adsorption and 'antibiotic-free' antibacterial activity. *Environmental technology*, 42(13), 2046–2066.

[12] El-Dib, F., et al. (2020). Advanced Industrial and Engineering Polymer Research.

[13] Fan, T. F., Park, S., Shi, Q., Zhang, X., Liu, Q., Song, Y., . . . & Cho, N. J. (2020). Transformation of hard pollen into soft matter. *Nature communications*, 11(1), 1449.

[14] Ghorpade, V. S., Dias, R. J., Mali, K. K., & Mulla, S. I. (2019). Citric acid crosslinked carboxymethylcellulose-polyvinyl alcohol hydrogel films for extended release of water soluble basic drugs. *Journal of Drug Delivery Science and Technology*, 52, 421–430.

[15] Goh, K. B., Li, H., & Lam, K. Y. (2018). Modeling the urea-actuated osmotic pressure response of urease-loaded hydrogel for osmotic urea biosensor. *Sensors and Actuators B: Chemical*, 268, 465–474.

[16] Goh, K. B., Li, H., & Lam, K. Y. (2019). Modeling the dual oxygen-and pH-stimulated response of hemoglobin-loaded polyampholyte hydrogel for oxygen-pH coupled biosensor platform. *Sensors and Actuators B: Chemical*, 286, 421–428.

[17] Guo, Y., Bae, J., Fang, Z., Li, P., Zhao, F., & Yu, G. (2020). Hydrogels and hydrogel-derived materials for energy and water sustainability. *Chemical Reviews*, 120(15), 7642–7707.

[18] Kaczmarek, B., Nadolna, K., & Owczarek, A. (2020). The physical and chemical properties of hydrogels based on natural polymers. Hydrogels Nat Polym, 151–172.

[19] Ke, Y., et al. (2018). Emerging thermal-responsive materials and integrated techniques targeting the energy-efficient smart window application. 28(22), 1800113.

[20] Uchida, M., Sengoku, T., Kaneko, Y., Okumura, D., Tanaka, H., & Ida, S. (2019). Evaluation of the effects of cross-linking and swelling on the mechanical behaviors of hydrogels using the digital image correlation method. *Soft Matter*, 15(16), 3389–3396.

[21] Zhang, D., Liu, Y., Liu, Y., Peng, Y., Tang, Y., Xiong, L., . . . & Zheng, J. (2021). A general crosslinker strategy to realize intrinsic frozen resistance of hydrogels. *Advanced Materials*, 33(42), 2104006.

[22] Anseth, K. S., Bowman, C. N., & Brannon-Peppas, L. (1996). Mechanical properties of hydrogels and their experimental determination. Biomaterials, 17(17), 1647–1657.

[23] Arslan Bin Riaz, M., Nasir, M. A., Nauman, S., Amin, S., Mehmood, F., Anwar, O. B., . . . & Abid, T. (2021). Passive cooling performance of polyacrylamide hydrogel on wooden and brick houses and effect of nanoparticle integration on its mechanical strength. *Plastics, Rubber and Composites*, 50(7), 340–350.

[24] Lee, J.-H., & Kim, H.-W. (2018). Emerging properties of hydrogels in tissue engineering. J Tissue Eng, 9, 2041731418768285.

[25] Tatiana, S., & Paiva, S., SCAFFOLDS DE HIDROGEL/FIBRAS ELECTROFIADAS COM GRADIENTES 3D PARA ORIENTAÇÃO CELULAR EM ENGENHARIA DE TECIDOS ÓSSEOS HYDROGEL/ELECTROSPUN SCAFFOLD WITH 3D.

[26] Lei, J., Li, Z., Xu, S., & Liu, Z. (2021). Recent advances of hydrogel network models for studies on mechanical behaviors. *Acta Mechanica Sinica*, 37, 367–386.

[27] Li, H., Luo, R., & Lam, K. Y. (2009). Multiphysics modeling of electrochemomechanically smart microgels responsive to coupled pH/electric stimuli. Macromol Biosci, 9(3): pp. 287–297.

[28] Nicolella, P. (2022). Engineering Dual Dynamic Polymer Networks with Tunable Elasticity and Diffusive Permeability (Doctoral dissertation, Johannes Gutenberg-Universität Mainz).

[29] Okay, O. (2009). General Properties of Hydrogels, in Hydrogel Sensors and Actuators (pp. 1–14). Springer Berlin, Heidelberg.

[30] Li, H., Ng, T. Y., Yew, Y. K., & Lam, K. Y. (2007). Meshless Modeling of pH-Sensitive Hydrogels Subjected to Coupled pH and Electric Field Stimuli: Young Modulus Effects and Case Studies. *Macromolecular Chemistry and Physics*, 208(10), 1137–1146.

[31] Liu, Q., Li, H., & Lam, K. (2018). Transition of magnetic field due to geometry of magneto-active elastomer microactuator with nonlinear deformation. J Microelectromechanical Syst, 27(2), 127–136.

[32] Liu, Q., Li, H., & Lam, K. (2017). Model development and numerical simulation of magnetic-sensitive hydrogels subject to an externally applied magnetic field. Procedia Eng, 214, 93–97.

[33] Liu, Q., Li, H., & Lam, K. (2017). Development of a multiphysics model to characterize the responsive behavior of magnetic-sensitive hydrogels with finite deformation. J Phys Chem B, 121(22), 5633–5646.

[34] Luo, R., Li, H., Birgersson, E., & Lam, K. Y. (2008). Modeling of electric-stimulus-responsive hydrogels immersed in different bathing solutions. *Journal of Biomedical Materials Research Part A: An Official Journal of The Society for Biomaterials, The Japanese Society for Biomaterials, and The Australian Society for Biomaterials and the Korean Society for Biomaterials*, 85(1), 248–257.

[35] Maihan, R. (2021). Uyarıya duyarlı mikrojel içeren amfoter hidrojellerin sentez ve karakterizasyonu= Synthesis and Characterization of Amphoteric Hydrogels Containing Stimuli-Responsive Microgels. ESOGÜ, Fen Bilimleri Enstitüsü.

[36] Liu, Z., Zheng, S., Li, Z., Xu, S., Lei, J. J., & Toh, W. (2022). Multiscale modeling of hydrogels. In The Mechanics of Hydrogels (pp. 187–222). Woodhead Publishing. (Elsevier Series in Mechanics of Advanced Materials).

Tahseen Arshad, Muhammad Pervaiz, Zhiduan Cai, Guibin Xu,
Shahid Ali Khan*

Chapter 3
Nanoengineering hydrogels with improved antimicrobial characteristics

Abstract: In recent years, hydrogels have secured significant consideration as a result of their extraordinary characteristics in the biomedical field. Researchers have been working on developing hydrogels of biopolymers with the incorporation of different bactericidal substances. Hydrogels possess characteristics such as biodegradability, biocompatibility, enhanced mechanical strength, and therapeutic properties, which made them suitable candidates for use in the domain of biomedicine. Hydrogels provide controlled drug release, wound healing, regulating cell proliferation and inflammatory response, and specific recognition of cell receptors, with improved mechanical strength similar to the extracellular matrix, essential for antibacterial biomaterials. The main emphasis of this section is on the medical applications of nanocomposite hydrogels, as well as the latest activities to overcome their current limitations. An emerging global public health issue is the quick spread of antibiotic resistance in pathogenic microorganisms. Antibiotics used locally might be the reason. Materials must function as the drug delivery method in local applications. The drug delivery mechanism should be biodegradable to meet clinical demand and offer sustained antibacterial activity. Due to its excellent hydrophilicity, unique three-dimensional structure, favorable biocompatibility, and ability to adhere to cells, a hydrogel is considered one of the most effective biomaterials for

Acknowledgments: The Authors highly acknowledge the Department of Chemistry, School of Natural Sciences, National University of Science and Technology (NUST), Islamabad 44000, Pakistan for providing necessary facility to this work.

*Corresponding author: Shahid Ali Khan, Department of Chemistry, School of Natural Sciences (SNS), National University of Sciences and Technology (NUST), Islamabad 44000, Pakistan; Department of Urology, Key Laboratory of Biological Targeting Diagnosis, Therapy and Rehabilitation of Guangdong Higher Education Institutes, The Fifth Affiliated Hospital of Guangzhou Medical University, Guangzhou Medical University, Guangzhou 510700, China, e-mail: shahid.ali@sns.nust.edu.pk

Tahseen Arshad, Department of Chemistry, School of Natural Sciences (SNS), National University of Sciences and Technology (NUST), Islamabad 44000, Pakistan

Muhammad Pervaiz, Department of Basic and Applied Chemistry, Faculty of Science and Technology, University of Central Punjab, Lahore 54000, Pakistan

Zhiduan Cai, Guibin Xu, Department of Urology, Key Laboratory of Biological Targeting Diagnosis, Therapy and Rehabilitation of Guangdong Higher Education Institutes, The Fifth Affiliated Hospital of Guangzhou Medical University, Guangzhou Medical University, Guangzhou 510700, China

https://doi.org/10.1515/9783111334080-003

drug delivery in the field of antimicrobial treatment. In order to address antibiotic resistance, metal nanoparticle-loaded hydrogels were emphasized.

Keywords: Hydrogels, Nanoscale materials, Antibacterial study

3.1 Hydrogels

Hydrophilic polymeric networks, known as hydrogels, are cross-linked to form a compact three-dimensional structure that remains water-insoluble. Through physical or chemical cross-linking, these polymeric hydrogels can maintain their structure while retaining an excess of 99.9% water content [1, 2]. Hydrogels possess remarkable attributes, including superior mechanical properties, high electrical conductivity, magnetic response, and antioxidation capabilities, which make them ideal for the production of nanocomposite hydrogels through advanced hydrogel synthesis techniques [3]. Hydrogels owe their water-absorbing capacity to specific hydrophilic functional groups. The significant water retention of hydrogels has gained the interest of researchers who seek to develop diverse approaches for potential biomedical applications, including but not limited to contact lenses, wound dressings, and drug delivery systems [4]. The smart response of polymeric hydrogels in several media like aqueous, electrolyte, pH, heat, light, pressure, DNA, RNA, and different solvents is basically due to changes in physical–chemical properties [5]. This swelling response in distinct media made them fascinating in various fields such as the pharmaceutical, agricultural, medical, chemical, and food industries [6]. Hydrogels that are sensitive to pH changes have demonstrated exceptional efficacy in various biomedical applications, particularly those of drug delivery systems [7].

3.2 Classification of hydrogels

Hydrogels can be categorized based on various parameters, including the type of cross-linking employed (i.e., physical or chemical cross-linking), different fabrication methods (i.e., homopolymer, copolymer, and interpenetrating networks), physical configuration (i.e., amorphous, crystalline, and semicrystalline), charge or presence of different types of functional groups (i.e., anionic, cationic, nonionic, and ampholytic), and on the base of origin (i.e., synthetic and natural hydrogels) [8].

3.2.1 Synthetic hydrogels

Synthetic polymers offer several characteristics including hydrophilicity, exceptional chemical stability, superior biodegradability, coating and gel developing tendency, and excellent mechanical properties. Synthetic polymers are used to synthesize for achieving enhanced mechanical properties [9]. While natural polymers possess many impressive attributes, their mechanical properties are often inadequate and need to be augmented with materials that can improve their strength. Synthetic polymers are utilized to enhance the mechanical and other desirable characteristics of natural polymers. Common examples of synthetic polymers include polyvinyl alcohol (PVA), polyacrylate, polyvinylpyrrolidone, polyacrylamide (PA), polyacrylic acid, polylactide, and polymethylmethacrylate [10]. Synthetic polymers are inherently nonbioactive and cannot be degraded easily. Cross-linking agents and initiators which are required during the synthesis process also cause additional cytotoxicity. Because of these factors, applications of synthetic polymers are greatly limited in the biomedical field.

3.2.2 Natural hydrogels

Naturally available materials such as collagen, cellulose, hyaluronic acid, alginate, gelatin, carboxymethyl cellulose (CMC), carboxymethyl chitosan, chitin, and chitosan (CS) have tremendous biocompatibility and structural resemblance to the extracellular matrix (ECM) [41], so they are considered encouraging materials for biomedical applications. Naturally available polymers are preferably utilized for biomedical applications, having superior properties and characteristics including cost-effectiveness, biocompatibility, and biodegradability as well as an effective biomaterial because they have biological molecular resemblance. When natural polymers are introduced into the human body, they can influence the behavior of immune systems.

The exclusive characteristics of natural polymeric hydrogels have tremendous applications in numerous fields including packaging, medicine, and biomedical fields like tissue engineering, biosensors, and bioimaging [11]. Natural polymeric hydrogels have excellent thermal and mechanical properties. The structural and chemical nature of CS provides bioactive intrinsic properties such as hemostatic, bacteriostatic, anticholesteremic, and anticarcinogenic [12]. Natural polymeric hydrogels have a significant character in the formulation of bioactive materials. Novel methodologies have been adopted to develop smart hydrogels, which exhibited excellent performance and gained tremendous attention in the biomedical field.

3.3 Properties of hydrogel

3.3.1 Biocompatibility

Biocompatible materials perform their functions in biological systems without leaving any harmful impact. Hydrogels are three-dimensional polymeric networks prepared by an appropriate material that has a resemblance to the ECM of soft tissues [13]. Hydrogels are biocompatible and lower the resistance of the guard cells of the immune system as hydrogels reached the targeted site and begin their function [14]. The adhesion of cells onto the hydrogels mainly depends on the functional groups present: carboxylic or acid salt groups provide minimum adhesion properties. The biocompatibility of hydrogels into biological systems can be improved by minimizing the response of immune cells which can be done by controlling the adhesion properties, especially in tissue engineering and scaffolding [15, 16]. The biocompatibility of hydrogels made them an excellent candidate for biomedical applications [17].

3.3.2 Biodegradability

In the field of biomedicine, hydrogel degradation plays a crucial role in various applications such as soft tissue engineering scaffold, surgical adhesions barriers, and delivery systems for medications. For instance, regulated hydrogel degradation in tissue engineering should encourage cell secretions and matrix remolding. But until the new tissue is merged with former at the implantation site, the hydrogel has to hold its shape. The hydrogel should break down and leave the body once it is no longer required. By using this technique, the hydrogel can be removed from the body without surgical treatment. Moreover, controlled hydrogel breakdown can help with the release of therapeutic drugs that have been loaded onto the hydrogel, particularly when delivering big molecules that are challenging to release through diffusion. Degradation rate is influenced by the rate of removal of drug [18]. As a result, it is crucial to adjust the deterioration rate in accordance with the intended use. A hydrogel typically degrades via one or more of the following mechanisms: hydrolysis of ester bonds, enzymatic hydrolysis, or photolysis cleavage. Hydrogels can be degraded by using different approaches, including the backbone of the polymer, cross-links, or pendant groups must be induced. In bulk degradation, entire hydrogel degrades, whereas in surface degradation, the degradation begins at the surface [19].

3.3.3 Mechanical properties

A hydrophilic three-dimensional polymer structure made of hydrogel contains functional groups that are accountable for water retention. The mechanical characteristics of hydrogels are affected by their capacity to swell. The higher ability of water absorption decreases the mechanical property of hydrogels. The mechanical features of hydrogels are also affected by the extent of cross-linking, as the intensity of hydrogels intensifies in proportion to the degree of cross-linking. Blending with other polymers via copolymerization or forming interpenetrating networks and forming a composite enhances the mechanical properties by inducing hydrophobicity into the hydrogel. Nanoparticles can also be incorporated into polymeric hydrogels to synthesize nanocomposites, which boost mechanical toughness by elevating the elastic modulus of hydrogels [20]. Moreover, the integration of nanomaterials into hydrogels intensifies the entanglement of the polymeric structure, which enhances mechanical durability. Augmented mechanical resilience is indispensable for hydrogels employed in medical sectors such as regenerative medicine, wound dressings, and ocular prosthetics [21]. The properties of hydrogels are summarized in Figure 3.1.

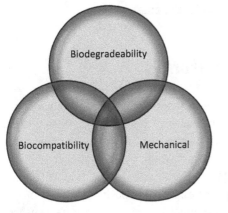

Figure 3.1: Properties of hydrogels.

3.4 Nanocomposite hydrogels

The polymeric hydrogels need further modifications, so these are reinforced with nanomaterials in order to enhance their properties, like mechanical strength, thermal properties, as well as swelling properties. After the dispersion of nanoparticles into the polymeric network, nanoparticles give new properties along with improved existing properties to the composite. Reinforcement of organic or inorganic nanoparticles such as carbon nanotubes (CNTs), GO, FeS_2, FeO, and MoO_3 into the polymeric hydrogels enhanced the mechanical properties [22]. The inorganic nanoparticles complemented with biopolymers also offered power, firmness, and extended utilization along with controlled drug release and other biomedical applications [23].

3.5 Reinforcement of nanoparticles

Polymeric blends possessed a mechanically weak assembly, which indicated their limited utilization in different fields. Different reinforced materials have been introduced into the polymeric network of hydrogel to improve their properties like swelling capacity and mechanical strength [24]. Nanocomposite hydrogels are cross-linked networks containing entrapped nanoparticles within the structure of hydrogel. Therapeutic nanocarriers have been demonstrated to be outstanding in traditional medical applications as a result of their remarkable characteristics in various domains, particularly in the area of administering drugs [25]. Nanoparticles that can be incorporated into the polymer blends include carbon-based nanoparticles (CNTs, graphene, and nanodiamonds), polymeric nanoparticles (dendrimers and highly branched polymers), inorganic/ceramic nanoparticles (hydroxyapatite, silicates, and carbon phosphates), and metal/metal oxide nanoparticles (gold, silver, and iron oxides). Among these nanomaterials, metallic nanoparticles exhibit excellent antimicrobial activities.

3.6 Metallic nanoparticles/hydrogels

Nanocomposite hydrogels containing both polymeric and metallic/metal oxide nanoparticles can be fabricated by incorporating nanoparticles such as silver (Ag), gold (Au), copper (Cu), zinc oxide (ZnO), and iron oxide (Fe_3O_4 and Fe_2O_3) into the polymeric hydrogels [26]. Nanocomposite hydrogels containing metal/metal oxide nanoparticles exhibit excellent properties including electrical, thermal, conductive, magnetic, and antimicrobial properties. Because of these properties, nanocomposite hydrogels incorporated with metal/metal oxide nanoparticles have been used in electronics, sensors, conductive scaffolds, and biomedical applications, especially in drug delivery systems. The biomedical applications of nanocomposite hydrogels is depicted in Figure 3.2.

3.7 The antimicrobial characteristics of metallic nanoparticles

Heavy metals exhibit antimicrobial properties, and since ancient times, metals like copper, gold, and silver have been utilized as antimicrobial agents. These metals possess antibacterial traits and are stable under physiological circumstances.

Polymer/metal nanocomposites can be produced through various methods, such as the synthesis of nanoparticles within a hydrogel or the addition of metal nanofillers directly into a hydrogel matrix. These approaches expand the antimicrobial potential of these metals. The antibacterial mechanism of these nanocomposites is based

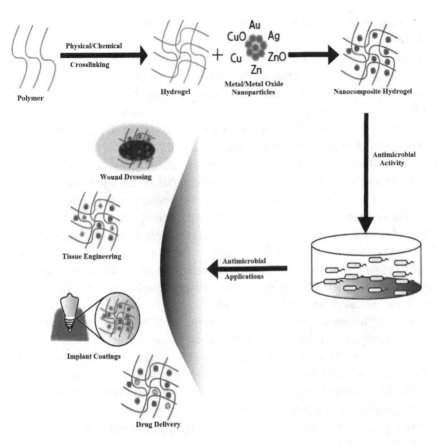

Figure 3.2: Antimicrobial nanocomposite hydrogels for biomedical applications.

on the stages that bacteria go through to be killed in a physiological environment. When water-containing dissolved oxygen and bacteria that have adhered to the polymer surface enter the polymer matrix from the surrounding media, the metal nanoparticles embedded within the composite begin to corrode upon contact with oxygen molecules. This corrosion releases metal ions that harm the bacterial membrane when they come into contact with the composite surface and subsequently infiltrate the bacteria.

3.7.1 Silver nanoparticles (Ag NPs)

Since before the term "microorganisms" was coined, silver has been regarded as an effective antibacterial agent. Food stuff was kept in their original conditions using utensils and other containers made of silver [27]. Hippocrates is credited with being the first to record the use of silver powder in the treatment of ulcers and wounds. In applications such as bone implants and wound dressings, silver continues to have a crucial role [28].

With advancement in nanoengineering and technology, silver is currently primarily used as nanoparticles. Antimicrobial silver nanoparticles (Ag NPs), being nontoxic, are harmless and exhibit a weak affinity for surface binding. These nanoparticles can form and develop in the open gaps between the cross-linked hydrogel networks when they are swollen. These areas actually serve as a nanoreactor. Ag NPs show antibacterial efficacy against a variety of microorganisms, including viruses, fungi, and drug-resistant bacteria, likely as a result of the many antimicrobial mechanisms they possess. Ag NPs demonstrate their efficacy as effective antibacterial agents due to their diverse sterilization methods. Recent research suggests that the primary mode of Ag NPs' antibacterial action is the discharge of Ag^+ ions. Ag NPs' particle-specific activity cannot be disregarded, which suggests that Ag^+ and Ag NPs have different modes of antibacterial action. The most commonly acknowledged theory is that Ag^+ discharged from Ag NPs interacts with cysteine at distinct protein sites situated on bacterial membranes, leading to a loss of internal K^+ and the failure of cellular transport mechanisms. Ultimately, this culminates in the demise of bacterial cells [29, 30].

Current research aims to create gel networks with Ag NPs. The antibacterial effectiveness of the hydrogel nanocomposite can also be influenced by physicochemical properties such as size, shape, surface area, surface chemistry, and oxidation state of Ag NPs. Because they could easily escape from the hydrogel network and interact with bacteria, the findings of the antibacterial activity test on bacteria demonstrate that smaller nanoparticles have better antibacterial activity while Ag NPs with larger sizes even though they can easily escape from the hydrogel network do not exhibit strong antibacterial activity.

3.7.2 Gold nanoparticles (Au NPs)

Although Au NPs have a variety of biological roles, gold is generally thought to be physiologically inert [31]. Au nanoparticles are considered to be biocompatible substances due to their ability to be synthesized in various dimensions and functionalized with preferred polymers [32]. If Au nanoparticles attach to the bacterial membrane, it could result in the leakage of bacterial contents or the penetration of the outer membrane along with the peptidoglycan layer, resulting in the eventual demise of the bacteria. As compared to investigations involving Ag NPs, studies conducted on antibacterial Au NP hydrogels are noteworthy. Recently, Brown et al. found Au NPs to be lacking in antibacterial properties. Nevertheless, Au NPs attached with ampicillin on their surface were found to be effective in destroying methicillin-resistant *Staphylococcus aureus*, *Pseudomonas aeruginosa*, *Enterobacter aerogenes*, and *Escherichia coli* K-12 substrain DH5-alpha (pPCR-Script AMP SK$^+$). Despite the prior reports referring to hydrogels based on *N*-isopropyl acrylamide having Au NPs and pH-responsive microcapsules of polymethacrylic acid hydrogel containing Au NPs, Gao et al. demonstrated that a hydrogel containing liposomes stabilized by Au NPs for antimicrobial application exhibited remarkable antibacterial properties against *S. aureus*, without causing any skin toxicity in a mouse model [33, 34].

3.7.3 Zinc oxide nanoparticles (ZnO NPs)

In addition to silver and gold, there are numerous additional metal nanoparticles with antibacterial properties, although only a small number are incorporated into hydrogels [35, 36]. Zinc is the most well-liked antibacterial agent of them. ZnO nanoparticles (ZnO NPs) are extensively utilized in various cosmetic products due to their renowned antibacterial properties and noncytotoxicity at a specific dosage. ZnO NPs use a variety of ways to fight microorganisms. ZnO NPs can firmly attach to bacterial cell membranes in several ways, causing the breakdown of the membrane's proteins and lipids. This process enhances the membrane's permeability and ultimately results in the lysis of the bacterial cell [36, 37]. In addition to harming the bacterial cell, ZnO NPs also produces reactive oxygen species such as hydrogen peroxide (H_2O_2) and Zn^{2+} ions. Owing to their ability to combat bacterial spores that are capable of withstanding high temperatures and pressures, ZnO NPs are effective against Gram-positive bacteria as well as Gram-negative bacteria. Like Ag NPs, ZnO NPs were integrated into the hydrogel network, resulting in the synthesis of antimicrobial hydrogel coatings. These coatings have demonstrated their potential as ideal contenders for latest biomedical device coatings [38, 39].

3.7.4 Metal nanoparticle-based antimicrobial hydrogels

Numerous metallic nanoparticles, which are combined with hydrogels, have been studied in the last few years. Besides these frequently used metallic nanoparticles, chitin hydrogels containing cytocompatible nickel nanoparticles were developed to fight against *S. aureus*. Furthermore, a natural zeolite/PVA hydrogel, infused with antibacterial cobalt, was shown to exhibit antibacterial properties against *E. coli*. Although not as potent as Ag NPs in terms of antibacterial properties, Cu NPs (which may consist of pure Cu NPs or CuO NPs) exhibit a broader range of microbicidal activity against various microorganisms, including fungi such as *Saccharomyces cerevisiae* as well as bacteria like *E. coli*, *Listeria monocytogenes*, and *S. aureus* [40]. In recent research, both copper-loaded CS hydrogel and CMC/CuO nanocomposite hydrogels exhibited favorable bactericidal effects against *E. coli* and *S. aureus*, while exhibiting no signs of toxicity. Additionally, there are reports indicating that nanoparticles made of magnesium, such as magnesium oxide nanoparticles (MgO NPs) and magnesium halide nanoparticles (MgX$_2$ NPs), use various approaches to impede bacterial proliferation. To explore drug administration in the digestive system, Hezaveh and Muhamad [41] utilized MgO NPs to create hydrogels by means of hydroxyalkyl-carrageenan derivatives. This could potentially suggest that the delivery of different drugs in a single system can be altered by incorporating metal nanoparticles or alternative materials into the hydrogel. Other metal nanoparticles, in contrast to Ag NPs and Au NPs, may require more research because they frequently appear in the design and production of contemporary medical biomaterials. The solution to antibiotic resistance might be a hydrogel with metal nanoparticles. These antibacterial polymers provide a number of benefits. First, metal nanoparticles may serve as an effective antibiotic replacement. Despite the universal usage of metal nanoparticles, reports of bacterial resistance remain uncommon. This is apparently because antibiotics typically only have one method of action, whereas antimicrobials have numerous mechanisms. Second, because of their small size, the particles can easily enter the cytoplasm of bacterial cells by slipping through cell membranes and peptidoglycan cell walls. Finally, metal nanoparticles are of stable quality, which implies that once they are liberated from dead cells, they can continue to kill further microbial cells. In this way, metal nanoparticles could produce a long-lasting antibacterial impact. Lastly, hydrogels can provide a local application delivery method. As the concentration of nanoparticles increases, the antibacterial property gets better. The infected site may have a high concentration of metal nanoparticles [26].

3.8 Applications in the biomedical field

Hydrogels composed of nanocomposites have found extensive use in diverse domains, including hygiene-related items, coating of implants, engineering of tissues, and dispensation of drugs. These hydrogels, particularly in terms of their antimicrobial and antifungal attributes, have been significantly utilized. Hydrogels are widely used in hygienic products including cosmetics; usually, skin is treated with cosmetic products, and hydrogels in cosmetics provide the physical–chemical barrier and protect the body from several different kinds of environmental influence. Hydrogels have been developed to enhance the pharmacological characteristics of molecules and boost their effectiveness while minimizing adverse effects. These are also modified to enhance the chemical reliability and aqueous dissolution of bioactive molecules. For the controlled release of drugs, systems based on biopolymers and cross-linking agents are the most appealing materials. Nanocomposite hydrogels have a pronounced perspective in the biomedical field. The active agents can also be coated or distributed into the nanocomposite hydrogels and found exclusive applications in the biomedical field.

3.9 Conclusion

Current developments in synthetic as well as natural hydrogels either have inherent antibacterial characteristics or serve as antibiotic transporters. Due to the inappropriate application of antibiotics and other antimicrobial drugs, an excess of drug-resistant bacteria have emerged. Consequently, hydrogels, which are biomaterials with antimicrobial properties, could present an alternative and adaptable approach to conventional antibiotic treatments. A wide range of hydrogels can offer significant benefits such as sustained as well as controlled release, improved biocompatibility, enhanced mechanical strength, local administration, as well as induced switch on–off release. Antimicrobial hydrogels have various applications, including wound coverings, coatings for the urinary tract, contact lenses, osteomyelitis treatment, catheter-related infections, and infections of gastrointestinal tract, overcoming intimidating challenges in conventional therapy. New antimicrobial biomaterials, their combination, and new strategies will open up exciting new possibilities for the future of infection control.

Antimicrobial gel components must be able to penetrate immune cells as well as kill pathogenic microorganisms inside in order to effectively treat microbial illnesses. In order to successfully cure infections and avoid the formation of biofilms, hydrogels containing antibiotics, peptides, antimicrobial polymers, as well as metal nanoparticles can release the antimicrobial agents over an extended period of time. Gels loaded with biodegradable antimicrobial polymers are more desirable than hydrogels encapsulating metal nanoparticles or antibiotics because antibiotics are more likely to grow drug resistance and because it's more challenging to reduce metal nanoparticles toxicity because of their inability to degrade. Nanocomposite hydrogels are cutting-edge biomaterials with potential uses in a variety of biological and pharmacological fields. The nanocomposite hydrogels possess better chemical, electrical, physical, as well as mechanical properties as compared to conventional polymeric hydrogels. Greater electrostatic interactions between the nanoparticles as well as polymer chains also improve the performance of nanocomposite materials.

3.10 Future challenges

However, all the advancements made in the area of hydrogels incorporated with nanomaterials, their continuous cytotoxicity in vivo issues, mechanical qualities, biodegradation followed by stimuli responsiveness are still out of control. Hence, future research may focus on mixing numerous components to create better hydrogels that are embedded with nanomaterials. Further investigation is necessary in relation to the functionalization of nanoparticle surfaces and the techniques employed to fabricate nanostructures, in order to generate hydrogels that are both environmentally friendly and durable.

References

[1]   Lv, Q., Wu, M., & Shen, Y. (2019). Enhanced swelling ratio and water retention capacity for novel super-absorbent hydrogel. Colloids Surf A: Physicochem Eng Asp, 583, 123972.

[2]   Guo, Y., Bae, J., Fang, Z., Li, P., Zhao, F., & Yu, G. (2020). Hydrogels and hydrogel-derived materials for energy and water sustainability. Chem Rev, 120(15), 7642–7707.

[3]   Kaczmarek, B., Nadolna, K., & Owczarek, A. (2020). The physical and chemical properties of Hydrogels Based on Natural Polymers, Chapter-6, 151–172.

[4]   Lavrador, P., Esteves, M. R., Gaspar, V. M., & Mano, J. F. (2021). Stimuli-responsive nanocomposite hydrogels for biomedical applications. Adv Funct Mater, 31(8), 2005941.

[5]   Sikdar, P., Uddin, M. M., Dip, T. M., Islam, S., Hoque, M. S., Dhar, A. K., et al. (2021). Recent advances in the synthesis of smart hydrogels. Mater Adv, 2(14), 4532–4573.

[6]   Zhou, L., Jiao, X., Liu, S., Hao, M., Cheng, S., Zhang, P., et al. (2020). Functional DNA-based hydrogel intelligent materials for biomedical applications. J Mater Chem B, 8(10), 1991–2009.

[7]   Bernhard, S., & Tibbitt, M. W. (2021). Supramolecular engineering of hydrogels for drug delivery. Adv Drug Deliv Rev, 171, 240–256.

[8]   Ahmed, E. M. (2015). Hydrogel: Preparation, characterization, and applications: A review. J Adv Res, 6(2), 105–121.

[9]   Madduma-Bandarage, U. S., & Madihally, S. V. (2021). Synthetic hydrogels: Synthesis, novel trends, and applications. J Appl Polym Sci, 138(19), 50376.

[10]   Teodorescu, M., Bercea, M., & Morariu, S. (2019). Biomaterials of PVA and PVP in medical and pharmaceutical applications: Perspectives and challenges. Biotechnol Adv, 37(1), 109–131.

[11]   Elkhoury, K., Morsink, M., Sanchez-Gonzalez, L., Kahn, C., Tamayol, A., & Arab-Tehrany, E. (2021). Biofabrication of natural hydrogels for cardiac, neural, and bone tissue engineering Applications. Bioact Mater, 6(11), 3904–3923.

[12]   Tian, B., Liu, Y., & Liu, J. (2021). Chitosan-based nanoscale and non-nanoscale delivery systems for anticancer drugs: A review. Eur Polym J, 154, 110533.

[13]   Zhao, Z., Vizetto-Duarte, C., Moay, Z. K., Setyawati, M. I., Rakshit, M., Kathawala, M. H., et al. (2020). Composite hydrogels in three-dimensional in vitro models. Front Bioeng Biotechnol, 8, 611.

[14]   Balitaan, J. N. I., Hsiao, C. -D., Yeh, J. -M., & Santiago, K. S. (2020). Innovation inspired by nature: Biocompatible self-healing injectable hydrogels based on modified-β-chitin for wound healing. Int J Biol Macromol, 162, 723–736.

[15]   Zhao, Y., Song, S., Ren, X., Zhang, J., Lin, Q., & Zhao, Y. (2022). Supramolecular adhesive hydrogels for tissue engineering applications. Chem Rev, 122(6), 5604–5640.

[16]   Selvan, N. K., Shanmugarajan, T., & Uppuluri, V. N. V. A. (2020). Hydrogel based scaffolding polymeric biomaterials: Approaches towards skin tissue regeneration. J Drug Deliv Sci Technol, 55, 101456.

[17]   Kesharwani, P., Bisht, A., Alexander, A., Dave, V., & Sharma, S. (2021). Biomedical applications of hydrogels in drug delivery system: An update. J Drug Deliv Sci Technol, 66, 102914.

[18]   Cao, H., Duan, L., Zhang, Y., Cao, J., & Zhang, K. (2021). Current hydrogel advances in physicochemical and biological response-driven biomedical application diversity. Signal Transduct Target Ther, 6(1), 426.

[19]   Pan, J., Jin, Y., Lai, S., Shi, L., Fan, W., & Shen, Y. (2019). An antibacterial hydrogel with desirable mechanical, self-healing and recyclable properties based on triple-physical crosslinking. J Chem Eng, 370, 1228–1238.

[20]   Zaragoza, J., Fukuoka, S., Kraus, M., Thomin, J., & Asuri, P. (2018). Exploring the role of nanoparticles in enhancing mechanical properties of hydrogel nanocomposites. Nanomaterials, 8(11), 882.

[21]   Li, Z., Cheng, H., Ke, L., Liu, M., Wang, C. G., Jun Loh, X., et al. (2021). Recent advances in new copolymer hydrogel-formed contact lenses for ophthalmic drug delivery. ChemNanoMat, 7(6), 564–579.

[22] Yadollahi, M., Gholamali, I., Namazi, H., & Aghazadeh, M. (2015). Synthesis and characterization of antibacterial carboxymethyl cellulose/ZnO nanocomposite hydrogels. Int J Biol Macromol, 74, 136–141.

[23] Yadollahi, M., Gholamali, I., Namazi, H., & Aghazadeh, M. (2015). Synthesis and characterization of antibacterial carboxymethyl cellulose/CuO bio-nanocomposite hydrogels. Int J Biol Macromol, 73, 109–114.

[24] Gholamali, I., Asnaashariisfahani, M., & Alipour, E. (2020). Silver nanoparticles incorporated in pH-sensitive nanocomposite hydrogels based on carboxymethyl chitosan-poly (vinyl alcohol) for use in a drug delivery system. Regen Eng Transl Med, 6, 138–153.

[25] Asadi-Ojaee, S. S., Mirabi, A., Rad, A. S., Movaghgharnezhad, S., & Hallajian, S. (2019). Removal of Bismuth (III) ions from water solution using a cellulose-based nanocomposite: A detailed study by DFT and experimental insights. J Mol Liq, 295, 111723.

[26] Yang, K., Han, Q., Chen, B., Zheng, Y., Zhang, K., Li, Q., et al. (2018). Antimicrobial hydrogels: Promising materials for medical application. Int J Nanomedicine, 13, 2217.

[27] García-Barrasa, J., López-de-Luzuriaga, J. M., & Monge, M. (2011). Silver nanoparticles: Synthesis through chemical methods in solution and biomedical applications. Cent Eur J Chem, 9, 7–19.

[28] Hu, R., Li, G., Jiang, Y., Zhang, Y., Zou, J. -J., Wang, L., et al. (2013). Silver-zwitterion organic–inorganic nanocomposite with antimicrobial and antiadhesive capabilities. Langmuir, 29(11), 3773–3779.

[29] Taglietti, A., Diaz Fernandez, Y. A., Amato, E., Cucca, L., Dacarro, G., Grisoli, P., et al. (2012). Antibacterial activity of glutathione-coated silver nanoparticles against gram positive and gram negative bacteria. Langmuir, 28(21), 8140–8148.

[30] Guo, L., Yuan, W., Lu, Z., & Li, C. M. (2013). Polymer/nanosilver composite coatings for antibacterial applications. Colloids Surf A: Physicochem Eng Asp, 439, 69–83.

[31] Faoucher, E., Nativo, P., Black, K., Claridge, J. B., Gass, M., Romani, S., et al. (2009). In situ preparation of network forming gold nanoparticles in agarose hydrogels. Chem Comm, 43, 6661–6663.

[32] Daniel-da-Silva, A. L., Salgueiro, A. M., & Trindade, T. (2013). Effects of Au nanoparticles on thermoresponsive genipin-crosslinked gelatin hydrogels. Gold Bull, 46, 25–33.

[33] Guiney, L. M., Agnello, A. D., Thomas, J. C., Takatori, K., & Flynn, N. T. (2009). Thermoresponsive behavior of charged N-isopropylacrylamide-based hydrogels containing gold nanostructures. Colloid Polym Sci, 287, 601–608.

[34] Kozlovskaya, V., Kharlampieva, E., Chang, S., Muhlbauer, R., & Tsukruk, V. V. (2009). pH-responsive layered hydrogel microcapsules as gold nanoreactors. Chem Mater, 21(10), 2158–2167.

[35] Weir, E., Lawlor, A., Whelan, A., & Regan, F. (2008). The use of nanoparticles in anti-microbial materials and their characterization. Analyst, 133(7), 835–845.

[36] Hajipour, M. J., Fromm, K. M., Ashkarran, A. A., de Aberasturi, D. J., de Larramendi, I. R., Rojo, T., et al. (2012). Antibacterial properties of nanoparticles. Trends Biotechnol, 30(10), 499–511.

[37] Pelgrift, R. Y., & Friedman, A. J. (2013). Nanotechnology as a therapeutic tool to combat microbial resistance. Adv Drug Deliv Rev, 65(13–14), 1803–1815.

[38] Schwartz, V. B., Thétiot, F., Ritz, S., Pütz, S., Choritz, L., Lappas, A., et al. (2012). Antibacterial surface coatings from zinc oxide nanoparticles embedded in poly (n-isopropylacrylamide) hydrogel surface layers. Adv Funct Mater, 22(11), 2376–2386.

[39] Mohandas, A., PT, S. K., Raja, B., Lakshmanan, V. -K., & Jayakumar, R. (2015). Exploration of alginate hydrogel/nano zinc oxide composite bandages for infected wounds. Int J Nanomed, 10(sup2), 53–66.

[40] Kruk, T., Szczepanowicz, K., Stefańska, J., Socha, R. P., & Warszyński, P. (2015). Synthesis and antimicrobial activity of monodisperse copper nanoparticles. Colloids Surf B: Biointerfaces, 128, 17–22.

[41] Hezaveh, H., & Muhamad, I. I. (2012). Impact of metal oxide nanoparticles on oral release properties of pH-sensitive hydrogel nanocomposites. Int J Biol Macromol, 50(5), 1334–1340.

Saurabh Shekhar, Shailendra Bhatt, Rohit Dutt, Manish Kumar,
Rupesh K. Gautam*

Chapter 4
Stimulus-responsive hydrogel for tissue engineering

Abstract: Tissue engineering is steadily evolving and at the same time awakening the interest among scientific community as a tool to solve numerous medical problems like organ transplant failure, disease treatment, genetic disorder remediation, and stem cell culture. But tissue engineering has been plagued by drawbacks such as lack of suitable tissue support matrices which would support engineered tissues to grow in an in vitro condition. Hydrogels being a polymer have been generating interest to be used as support matrices in various tissue engineering processes due to its malleability properties. Hydrogels can be considered as frontrunners' cell support materials used in tissue engineering. Rigid structure of hydrogel provides ample support to growing cells and its inertness also ensures there is no interaction between cells and the polymer itself. Ample availability of variants of hydrogels also ensures its application in the fields of cell culture, self-healing, and drug delivery. Its ability to be used in different fields leads hydrogels to be called as smart hydrogels as they also find their use in soft robotics as well as amenable and modifiable accoutrements which can be used as bioenzymes, chemical sensors, and absorbents for carbon prisoner. The need of inexhaustible source of transplant organs also puts emphasis on utilization of hydrogels as scaffolds and as cell support material.

Keywords: Hydrogel, tissue engineering, scaffold, support matrices, tissue regeneration

4.1 Tissue engineering: a brief introduction

Tissue engineering (TE) has long been widely recognized and widespread in science and medicine (Organizer University of California, San Diego). Later, as the proposals made in 1985 failed to attract the attention of the scientific community, in 1987 Feng

*Corresponding author: Rupesh K. Gautam,** Department of Pharmacology, Indore Institute of Pharmacy, Rau, Indore, Madhya Pradesh, India, ORCID: 0000-0001-5580-5410,
e-mail: drrupeshgautam@gmail.com
Saurabh Shekhar, Shailendra Bhatt, Department of Pharmacy, School of Medical and Allied Sciences, GD Goenka University, Sohna Gurgaon Road, Sohna, Haryana, India
Rohit Dutt, Gandhi Memorial National PG College, Ambala Cantt, Haryana, India
Manish Kumar, MM College of Pharmacy, MM (Deemed to be University), Mullana-Ambala, Haryana, India

https://doi.org/10.1515/9783111334080-004

Yuancheng reinvented TE as a new field that could be the future of National Scholastic Athletics Foundation's "Engineering Bioengineering and Research Assistance Directorate for People with Disabilities." The scientific community is finally combining bioengineering and medical science to achieve major advances in TE.

In 1993, Robert Langer and Joseph Vacanti [1] received a grant from the National Science Foundation to publish their research focusing on the knowledge of new resources.

From the early 1990s to the year 2000, the number of papers on TE has expanded dramatically. With increase in advancements in technology in bioengineering and medical field during 2000s, TE as a means of biomedical engineering discipline has been emerging at a rapid rate and has become a big area of interest in both medical and engineering fields [2]. At present TE has become a multidisciplinary field that incorporates basics of life sciences, cell, and molecular biology. The TE methodologies depend upon matrices which would enable the cellular structure to get structured itself in the body or an artificial scaffold is used to maintain them into a specific structure as structural integrity of cells is of utmost priority.

The four basic aspects of TE are
1. cell sources and culture,
2. cell acclimatization,
3. cell-bearing materials, and
4. tissue preparation and development.

TE is the process of using cell culture models to create tissues/organs. The main goal of TE is to use various substances in the body to repair or replace distorted tissues. TE can be considered the mainstay of reorganization surgeries. With the development of TE, skin, nerves, ligaments, valves of heart, joints, and nervous tissue implants can be provided [3, 4]:
– Tissue production and engineering
– Cell history and culture
– Cell-bearing materials

The significance of TE is to stimulate regions of body to replace the distorted tissue. TE is often considered the cornerstone of reconstructive surgery [5].

There are two ways by which TE is realized:
(i) Some scientists believe that cells have the ability to produce fat. This means that when cells grow on a suitable support, the cells will proliferate and eventually form the functional tissue. The structure and functioning of the organization are the same as the old organization. This strategy is simple and cost-effective, but its results are limited.

According to other systems thinking, new business organizations emerge from the management process. Therefore, in vivo tissue regeneration and in vitro tissue synthesis are quite difficult. TE is more than just regenerating cells; it requires a holistic strategy that includes detailed analysis of the cell structure and special preparation and control procedures.

(ii) Autologous cell source: When a patient's brain is used for TE, the cell source is considered autologous. This is an easy concept to understand. A biopsy is used to obtain the desired tissue sample. It can be enzymatically digested or implanted, and cells can be expanded to a large size.

The following are the key advantages of autologous cells in TE:
1. Avoidance of immunological problems
2. Reduced risk of transmission of intrinsic infections

There are a few downsides to using autologous cells:
–   Obtaining sufficient biopsy material from the patient is not always achievable.
–   The disease status and age of the patient will be limited.

Allogeneic cell: Cells are taken from individuals except patient, and then the cells are termed allogeneic cell. Advantages of allogenic cells over other stem cells are:
–   Get more from healthy donors.
–   Cells will be enlarged.
–   It is cost-effective and durable.

The main problem with allogeneic cell sources is the immune system, which ultimately leads to transplant rejection. Role of immunity diverses with different types of allogenic cells utilized. For instance, endothelial stem cells represent higher resistance for antibiotics whereas smooth and fibroblast cells exhibit meager resistance. Disease prevalence is also affected by age factor in patients. Henceforth, adult donor stem cells tend to be more immunogenic as compared to pregnant or newborn exhibiting fewer instances of resistances [6]. When cells are taken from different species (such as pigs to humans), the source is called xenogenic. This method is not used often due to the immune system. Cell culture methods appropriation for TE relies on cells types and their function. Mostly, monolayer cultures preferred for TE but there are linked disadvantages such as losing of morphology, longevity, and utility as functional cells. It is preferred to use cells in three-dimensional culture for TE [7]. Food and fuel exchange are limited features of three-dimensional culture. Gene therapy has been used successfully in TE. Introducing the required gene into the culture is one of the methods to achieve gene therapy. New genes can be added to the combination of existing proteins or new proteins can be made. There are some successes in this direction:
A. Fibroblasts genetically modified to produce factor VIII and IX.
B. Altered endothelial tissue can produce cathepsin.

Keratinocytes genetically modified to produce trans-glutaminase-1 (patients with skin diseases and lamellar ichthyosis do not have this enzyme). Keratinocytes transformed with required genes editing have been proposed to be successful to transplanted rat models having diseases.

Cell acclimatization: In terms of quantity and surroundings, the following factors affect cell acclimatization:

A. Chemical constituents
B. Mechanical hints
C. Substrate orientation

The properties of the soil sublayer determine the direction of contact. These features are also ridge patterns, aligned fibers, and so on. Differentiation of the substrate can be used as a way to generate cells differently.

Synthetic matrix fibers made of collagen and fibronectin find their application as TE templates. Directing the cell and stimulating cell function requires the creation of chemical changes. Some growth factors and extracellular macromolecules can form chemical gradients, such as vascular endothelial protein, hyaluronic acid oligosaccharide fragments, and collagen fibers [8]. These are few aberrations in continuing optimum exchanges between cell cultures. This can become limited when cells become dense. The cellular response to radiation is rather complicated and is given as follows:

A. Cell variations
B. Cytoskeletal destortion
C. Change the matrices design
D. Regulates the cell cytology (e.g., growth factors and hormones)

Three main types of mechanical factors affecting cell cytology:

A. Drumming
B. Compression force
C. Cutting force

With increase in advancements in technology in bioengineering and medical field during 2000s, TE as a means of biomedical engineering discipline has been emerging at a rapid rate and has become a big area of interest in both medical and engineering fields. At present TE has become a multidisciplinary field that incorporates basics of life sciences, cell, and molecular biology. The TE methodologies depend upon matrices which would enable the cellular structure to get structured itself in the body or an artificial scaffold is used to maintain them into a specific structure as structural integrity of cells is of utmost priority [9].

The principles of TE include embryonic stem cells (ESCs) based on bone marrow-associated mesenchymal stem cells and Cemento-osseous dysplasia (COD)-associated mesenchymal stem cells. Embryonic cells can be stored for long periods of time, which is beneficial for their applications in TE. It also provides tissue compatibility for all patients. Stem cell storage provides a convenient method for medical cloning.

This makes stem cells an important candidate for TE. The pluripotent property of stem cells allows a large capacity for differentiation necessary for tissue formation. Pluripotency allows stem cells to tissue into multiple tissues with multipotency. The bone marrow-derived mesenchymal stem cell (BM-MSC) method is a method that can detect characteristics of bone, especially mineral deposits. This approach further enhances TE by allowing understanding of the differentiation history of stem cells. Mesenchymal stem cells of bone marrow are obtained from osteogenic cells which later realize tissue repair process. Bone marrow contains a mixture of many cells that is utilized in TE due to their high tendency to adapt and differentiate into different tissues. Artificial hip fillers used to provide stability can be used as a result of TE supported by BM-MSCs. Mesenchymal stem cells of umbilical cord are also highly differentiated, making them another candidate for TE. The gene-level resemblance between mesenchymal stem cells obtained from the umbilical cord and stem cells of mesenchymal origin from the bone marrow is worth noting. They are similar to genetics, thus suggesting their use in TE [10]. Mesenchymal stem cells derived from the umbilical cord differentiate to liver cells, adipocytes, osteoblasts [11], and neuron-like cells [12]. The biggest challenge in achieving TE is the availability of stem cells. Efficacy and efficiency will enable TE to further advance medical science in the treatment and first-line treatment of many diseases. The similarities between umbilical cord-associated mesenchymal stem cells and bone marrow from bone marrow and development of TE therapy, which will increase connective tissue for healing. This type of stem cell may provide a broad source of material that can be developed by noninvasive methods.

TE by definition involves large cluster of cells designed to differentiate into uniform type of cells performing similar function. These clusters of cells need to be enclosed inside the matrix to maintain shape which is an important aspect of TE developmental technique [13]. These matrixes should be mimicking the original tissue in their cluster formation. The TE success largely depends upon the adaptability of differentiating cells with the matrix employed. The cells obtained from the source need to be enclosed inside a suitable matrix of similar shape and conformity of the parent tissue. More the similarity between them the more will be the success rate in employed TE technique.

The introduction of TE scaffold inside the patient would always be a challenge as the cells' ability to differentiate is high resulting into high chances of rejection [14, 15]. On top of that, utilization of invasive methods also poses its own set of drawbacks. All these concerns have led to the integration such as ESCs, mesenchymal stem cells from bone marrow, and umbilical cord-linked mesenchymal stem cells, making them more popular TE tools.

The development of TE weighs on the type of scaffolds and matrices used to support different cells. If these scaffolds are to be used in TE, inertness is one of the sought-after properties. This allows the use of various polymers as scaffolds and matrix in TE. Biodegradable polymers, hydrogels in the form of polylactic acid, certainly represent candidates for TE. The polymers are similar and can be easily converted into any form and shape required for TE [16].

Customized scaffolds can further be tailored by plasma polymerization processes. There are many polymers used as scaffolds in TE. Poly-hydroxyl acids and poly lactic-*co*-glycolic are polymers which have been widely used in TE as scaffold. They have their advantages over other prevalent polymers as these are hydrolyzable which means sustained and controlled drug release for various diseases and also supports controlled degradation profile which allows them to be a perfect candidate for TE applications.

Magnetic hydrogels (iron oxide and hydrogel matrices) is employed in TE in biomedical applications (regenerative medicine that repairs damaged body tissue). Due to their biocompatibility, controllable structures, and "smart response to magnetic field remotely," they appear to be a suitable substance, indicating that they are biosensors that can be controlled remotely via the Smart Grid. The role of magnetic hydrogels in TE as scaffold is given in Figure 4.1.

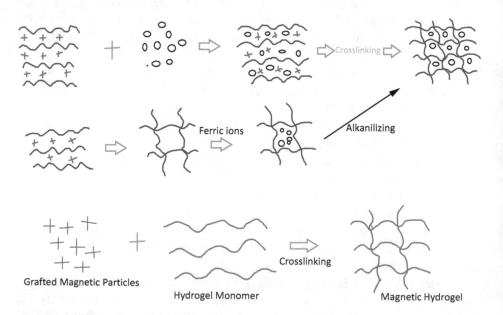

Figure 4.1: Magnetic-sensitive hydrogels in tissue engineering.

Apart from the abovementioned candidates, the use of high-pressure CO_2 foamed scaffolds, injectable scaffolds, and novel custom scaffolds has also been on the rise given their malleability and ease to adapt to desired conformation. All these materials can be utilized as scaffold in TE keeping in mind the molecular weight.

4.2 Hydrogels: a brief introduction

Hydrogels are water-loving three-dimensional mesh possessing ability to ingress high quantity of water and bodily fluids. This proposes them as tools for biosensors, cellular delivery agents, and tissue like matrices. This review sheds light on the benefits of hydrogels as compared to other biomaterials. Hydrogels respond to different stimuli given their varying chemical composition. They can respond toward heat, light, pH, and other chemicals. There are two proposed swelling mechanisms to understand the structural features of hydrogels followed by gel processing and lattice arrangement. Polysaccharide, peptide, and many synthetic hydrogels have been the horcrux of this review. In this chapter, biomedical application of hydrogels involving self-healing and drug delivery has also been pondered. As the name suggests, hydrogels are made of special materials that can hold large amounts of water. Its properties include different polymers held together by different bonds, giving it a unique three-dimensional shape. They have many applications including biosensors, drug delivery vehicles, and cell and matrix carriers in TE. Given chemical constitution and response to heat, pH, light, and chemicals, hydrogels have properties similar to other biomaterials.

Connections between network chains help prevent fragmentation. Hydrogels are a broad category that includes many cosmetic and artificial materials. Natural hydrogels involve biomaterials such as collagen and silk. Other biomaterials such as hyaluronic acid, alginate, chitosan, and fibroin too constitute the formation of natural hydrogel. Biosimilarity, biodegradability, reduced cytotoxicity, and ability to transform hydrogels into intravenous gels are distinguishing features of hydrogels. However, natural hydrogels exhibit several drawbacks such as below par mechanical properties and uncontrollable quality due to batch-to-batch variations [17, 18].

For these reasons, natural hydrogels are often mixed with synthetic fibers to create composite polymers. Hydrogels are extensively used in biochemistry given tunable packing, biocompatibility, and better compatibility with aqueous and cellular environments. Over the years, hydrogels have evolved from static materials to "smart" materials which respond to colored stimuli like pH, temperature, chemicals, electricity, and light [19]. This makes it useful for biomedical research as well as programmable materials such as soft robots and electronics, catalysts, chemical sensors, and carbon monoxide absorbers.

Hydrogels vary from other biomaterials by their ability to retain high water, swelling, and biosimilarity, which vouches for their biomedical applications. Due to their chemical structure and cross-linking, hydrogels can respond to heat, pH, light, and chemicals, which make them apt for many utilities. To better understand how structure size affects the performance of hydrogels in certain conditions, two alternative hydrogel swelling mechanisms are discussed. The classification and associated functionality of hydrogels is provided in the flowchart 4.1.

Further research should focus on various activities aimed at achieving strong mechanical, rapid, and self-healing properties and various biomedical applications.

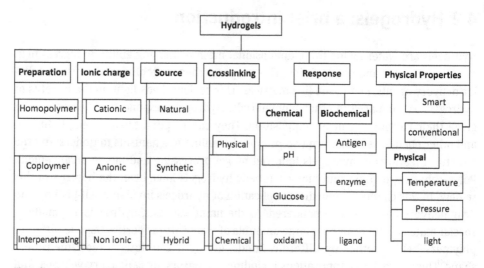

Flowchart 4.1: Classification and associated functionality of hydrogels.

4.2.1 Role of hydrogels in tissue engineering

The support material for cells often determines the properties of TE. The role of hydrogel in TE is provided in Fig. 4.2. The supporting materials can be broadly distinguished as follows:
A. Nonchemical materials
B. Bioprosthetic materials
C. Synthetic products
D. Natural polymers
E. Seminatural products

4.2.2 Nonchemical materials

Nonchemical materials like ceramics, metals, and plastics do not get absorbed into tissues or biointegrated. Therefore, it is better not to use conventional materials in TE [20].

4.2.3 Bioprosthetic materials

Natural materials modified for bioinert represent bioprosthetic materials. For example, normal collagenous connective tissue (e.g., porcine heart) can be stabilized by treatment with glutaraldehyde [21]. They are not considered immunogenic and do not undergo any modification even after years of transplantation. However, some tissues can grow on bioprosthetic materials.

4.2.4 Synthetic materials

As support materials, a variety of synthetic bioresorbable polymers are available. Poly(glycolic acid), poly(lactic acid), and the copolymer poly(lactic-*co*-glycolic acid) are the three polymers most frequently utilized in TE. In addition to promoting cell proliferation in vivo, modifications have also been made to stabilize cells in vivo [22]. Using synthetic materials has several advantages. It is easy and cheap to make. The polymer composition is also reproduced in large-scale production. However, it also has disadvantages:
a. Not compatible with natural polymer-based cells
b. Decomposition produces substances that can damage cells

4.2.5 Natural polymers

The most commonly used synthetic material is collagen-chondroitin sulfate granules. It is available under the brand name Integra and is available in various formulations such as collagen gel, fibrin gel, Matrigel, and some polysaccharides. These contain chitosan and hyaluronic acid as a polysaccharide moisturizing gel.

Natural polymers promote close packing of molecules using only the principles of interaction in polymeric materials. Molecules formed in this way can be used as supporting materials.

Listed below are some examples of semiactive products:
A. Hyaluronic acid that has been chemically cross-linked and stabilized via benzyl esterification
B. Tannins or carbodiimides induced collagen cross-linked products

Organizational design and engineering:
A. The work needs to be corrected immediately.
B. It is easy to repair the tissue.
C. The patient is least comfortable.

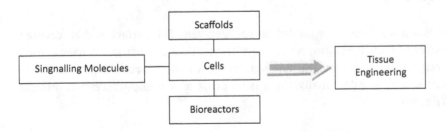

Figure 4.2: Role of hydrogels in tissue engineering.

Sources of donor cells have been widely criticized from the point of view of TE. To suppress the immune system, it is better to use the patient's cells (autologous cells). Allogeneic tissues are utilized especially in temporary treatment. The antigenicity of allogeneic cells has been shown to decrease when the cells are cultured and/or stored (e.g., cryopreserved). Other important factors in TE are support matrices and cell attachment properties.

4.2.6 Dermatology enhancement by TE

Human keratinocytes could be cultivated in a lab setting in a way that was appropriate for transplantation for the first time in 1975. Epithelial cells grew to form a continuous layer and merged into tissue layers [23]. TE is the main part of dermis, which contains blood vessels, nerves, sweat glands, and other internal organs. In the recent years, there have been advances in the creation of skin substitutes that can be considered TE skin structures.

4.2.7 Integra™

Integra™ is a collagen and glycosaminoglycan consisting biosynthetic material. It is not considered a real TE model and utilized only for transplantation of transplanted cells.

4.2.8 Dermagraft™

It consists of a poly(glycolic acid) polymer network added to human dermal fibroblast stem cells that are derived from the foreskin of infants.

4.2.9 ApliGraph™

Human skin fibroblasts were seeded on collagen gels. The surface is then covered with a layer of human keratinocytes. The aforementioned paper goods have a shelf life of roughly 5 days. They might, however, merge with the surrounding tissue to create wholesome skin. Additionally, there is no proof of immunosuppression brought on by TE creation.

4.3 Advantages associated with the use of hydrogel in tissue engineering

Many utilizations of hydrogels in TE which can be attributed to the fact that hydrogels can be found in various forms. There are many stimuli-responsive hydrogels that can be used for TE. According to its properties, hydrogel provides multifaceted approach for achieving tissue engineering successfully.

Depending on the application, hydrogels are formulated to react to a variety of bodily stimulus like pH, ion balance, and heat.

4.3.1 Hydrogels with thermo-sensitive properties

The key to controlling the properties of thermo-sensitive hydrogels is to maintain a balance between hydrophobic and hydrophilic domains. Temperature has an impact on both hydrophobic interactions between hydrophobic polymer segments and water molecules as well as hydrophilic interactions between hydrophilic polymer segments and water molecules. Therefore, a slight change in temperature can disrupt the original equilibrium and lead to a sol-gel transition. Long and colleagues [24] studied the use of poly (N-isopropylacrylamide) hydrogel sheets in smart thermochromic windows.

Solar control capacity (T_{sol}) is very high and light (luminous) transmittance (T_{lum}) is the highest value so far [25]. This hydrogel demonstrates a very good combination of high pressure and enhanced solar energy, potentially paving the way for the creation of thermochromic smart windows made from this material.

pH-sensitive hydrogels are biomaterials that have good physical and chemical properties when exposed to appropriate pH. The polymer chain is linked by acid or base groups. At high pH, acidic groups are deprotonated, while at low pH, the groups begin to be protonated. Hydrogel swelling in aqueous solutions occurs due to the binding, dissociation, and attachment of different ions to the polymer chain. A pH-responsive hydrogel consisting of hydrophobic PMMA nanoparticles and polymethacrylic acid-grafted ethylene glycol (P(MMA-g-EG)) was created by Peppa's group [26]. Different molar ratios of PMMA nanoparticles were added to P(MMA-g-EG) to create an amphiphilic polymer with various physical characteristics. pH change from stomach to intestine was used to test the release of the encapsulated drug.

4.3.2 Hydrogels responsive to light and drugs

Photosensitive hydrogels are attractive materials in drug/gene delivery, microlenses, and sensing because photosensitive processes can be localized distantly and irrelevantly. When exposed to 355 nm light, the hydrogel, which is made of a deoxycholic acid-modified cyclodextrin derivative and an azobenzene-branched polyacrylic acid co-polymer, may successfully transition from the gel phase to the sol phase [27]. Irradiation with 450 nm light causes the hydrogel to change from the sol phase to the gel phase. The transformation of this hydrogel can be moderately controlled, which means it can be used in bioengineering applications to release molecular and cellular drugs.

4.3.3 Required for tissue engineering

4.3.3.1 Repair of functional matrix cells

Stem cells have the ability to differentiate to specialized and multiply themselves, but in some adults, they do not have movement differentiation. They are characterized as totipotent, multipotent, multipotent, or monopotent, according to their differentiation capacity or "potency." Totipotent stem cells can produce all cells in animals including placenta [28]. Pluripotent cells, also termed ESCs, can generate embryonic tissue from all three cell types.

Pluripotent stem cells can also be created by experiments from adult somatic cells in addition to innate pluripotent stem cells. By genetically reprogramming cells, "induced" pluripotent stem cells 1, 2 are created. Yamanaka and Gurdon received the Nobel Prize for their discovery in 2012, and the process was selected Nature's Method of the Year in 2009.

Pluripotent stem cells are cells that have lost the ability to produce any cell type. Multipotent cells can form a small number of cells in a cell family, such as all tissues in an organism or body. Adult or "somatic" stem cells are specialized cancer cells that specialize in causing many types of diseases in the body. Adult stem cells reside in specialized niches and interact with the local environment to maintain their roots. These stem cells help maintain tissue homeostasis and respond to injury [29].

4.3.3.2 Transplantations and support stents

The first kidney transplant occurred in 1954 in Boston, Massachusetts, when Joseph Murray and John Merrill and their colleagues successfully transplanted twins without the use of antibiotics. Kidney transplant procedures are used to treat chronic kidney failure and allow the body to transition from experimental to surgical treatment. Immunology research, the development of novel medications, and improvements in immunotherapy all have contributed to the advancements in transplantation. In addition, all these factors contribute to prolonging the life of the body and its receptors. Due to the benefits of organ transplantation, the demand for organ transplantation far exceeds the supply. Survivors, related and unrelated donors, transplants, and expanded kidney donation procedures are added to the donor pool. However, all this is not enough for those who expect change. Despite the use of immunosuppression, the mortality rate of dialysis patients who have been waiting for a transplant for a long time is higher than that of post-transplant patients. Considering all the above factors, a continuous supply of organs is urgently needed.

4.3.3.3 Kidney extracellular matrix

Intrinsic biocompatibility, design, and cardiovascular disease are the reasons why kidney extracellular matrix scaffolds have the ability to integrate into the recipient's body. This scaffold also has the ability to progress progenitor cells into a phenotype which is highly specific to an organ cells. Recent studies in TE and cell matrices transplantation have been fruitful. The idea behind cell scaffold seeding is that cells remain due to support, are seeded, and maintain their function. The extracellular matrix may play such a role in the in vitro scenario. The differentiated extracellular matrix structure preserves both organ's morphological shape and the renal parenchyma's macro- and microarchitecture. The biocompatibility often depends on the removal of critical reagents like human leukocyte antigen classes I and II [30]. There are still major concerns regarding tissue integrity when biocompatible materials are obtained from ineligible transplants. There are various studies exhibiting that such modifications are reversible.

4.3.4 Smart hydrogel benefits

Smart hydrogel is defined as a polymer network that may react to outside stimuli by abruptly changing the network's physical makeup:
A. Noninvasive
B. Remote-controlled therapies
C. Targeted drug delivery
D. Regenerative medicine
E. TE
F. Implanting artificial organs

4.4 Future perspectives in regenerative medicines

Regenerative medicine carries potential for revolutionizing treatment of numerous degenerative diseases, such as osteoarthritis and chronic obstructive pulmonary disease. Harnessing the body's innate ability of self-healing, reparative, and regenerative treatment strategies holds the promise of managing damaged tissue and reversing of many degenerative diseases, providing hope for patients who are currently incurable. Degeneration is the opposite of regeneration therapy. According to the National Institutes of Health, "aging is the process of working through tissue to repair or restore tissue or body function lost due to disease, injury, or disorder." Nutritional cell therapy is one of the many methods used in regenerative medicine.

As part of regenerative medicine, somatic cell treatment strategy has been applied to humans for many years. However, new developments in cellular biology have presented numerous possibilities to improve and develop established regenerative therapies. Stem cells have a tremendous ability to regenerate. Taking stem cells from one location in the body and transplanting them to the diseased region provides somatic cell therapy an impetus in harnessing the body's self-healing power. Stem cells' primary function is to repair and replenish the tissue where they reside. Somatic cell treatment strategy improves the self-healing process by directing progenitor cells to diseased region of the body. Adult stem cells are derived from a patient's tissues like fat, muscle, teeth, skin, or bone. Under anesthesia, stem cells are removed from the patient quickly and painlessly in approximately 30–90 min. After extraction, injection is usually done. The surgery is fair, legal, and causes no discomfort to the patient. Therefore, its many uses are virtually unlimited and include orthopedics, degenerative diseases, neurological, and autoimmune diseases.

4.4.1 Hybrid hydrogels in tissue engineering

4.4.1.1 Supercapacitor hydrogels

The rapid growth of wearable technologies, arbitrarily curved displays, and even transparent mobile phones necessitates the manufacturing of flexible, transparent, lightweight, and economical storage alternatives. The simultaneous inclusion of mechanical robustness, optical transmittance, and electronic conductivity is currently the most pressing challenge. Due to the necessity of high-performance flexible supercapacitors, supercapacitor technology continues to advance at a rapid pace.

4.4.1.2 Injectable hydrogels for spinal cord regeneration

Spinal cord injury (SCI) can be a difficult problem for regeneration due to the many characteristics of growth inhibition after SCI. Most of these diseases do not affect the meninges, and some axons remain undamaged and regenerate at the injured site. In this case, surgical implantation of a prefabricated frame or DDS into the spinal cord may cause pathology in the future [31]. Using scaffolds created in situ is an option for this technique. After injection into the injured area of the spinal cord, the viscoelastic hydrogel quickly transforms from liquid to gel and adapts to the tissue of the injured area.

4.4.1.3 Biosensor hydrogels

Biosensor hydrogels are inherently biosimilar given their high water content and hydrophilicity, making them similar to space-filling components of extracellular matrix. Therefore, one application of hydrogels in biosensors is to coat and cover sensor surfaces to prevent interaction with biomolecules or cells. The hydrogel will serve as a stable biosensing element by providing appropriate conditions for enzymes and other biomolecules to maintain their activity and structure.

4.5 Conclusion

TE is one of the new treatments. TE not only enables the treatment of diseases but also enables the production of anatomical materials (tissue/organ). The most difficult task is to create stable matrices for cell culture. The choice of matrix support depends on the ability of the matrix material to support cell growth and remain viable for the resulting cells. Hydrogels appear to have both of these properties and are therefore considered the supporting matrix material of choice for TE. The flexibility of hydrogels means they can be molded into any shape and structure, while being stiff enough to support cell growth makes them attractive scaffolding materials. Hydrogels are available in many forms, including chemical hydrogels and supercapacitor hydrogels, revealing new possibilities for future applications. Hydrogels such as collagen, fibroin, hyaluronic acid, chitosan, and alginate make it superior when compared to other materials. They are also known for their ability to progress from stem cells to phenotypes specific to a tissue type, such as the kidney extracellular matrix. As transplants for chronic kidney failure patients continue, the need for permanent stents is expected to increase. Despite the use of immunosuppression, the mortality rate of dialysis patients who have been waiting for a transplant for a long time is higher than that of post-transplant patients. Considering all the above information, organ transplantation should not be a possibility. Therefore, the demand for scaffolds has also increased, and hydrogels are important in TE as their plastics can be molded into various shapes and sizes and are strong enough to support TE cells.

Abbreviations

BM-MSC	Bone marrow-derived mesenchymal stem cell
ESC	Embryonic stem cell
P(MMA-*g*-EG))	Poly(methyacrylic-*graft*-ethylene glycol)
SCI	Spinal cord injury
TE	Tissue engineering

References

[1] Amiryaghoubi, N., Pesyan, N. N., Fathi, M., & Omidi, Y. (2020). Injectable thermosensitive hybrid hydrogel containing graphene oxide and chitosan as dental pulp stem cells scaffold for bone tissue engineering. Int J Biol Macromol, 162, 1338–1357. doi: 10.1016/j.ijbiomac.2020.06.138.

[2] Attalla, R., Ling, C. S., & Selvaganapathy, P. R. (2018). Silicon carbide nanoparticles as an effective bioadhesive to bond collagen containing composite gel layers for tissue engineering applications. J Adv Healthc Mater, 7, 1701385. doi: 10.1002/adhm.201701385.

[3] Augst, A. D., Kong, H. J., & Mooney, D. J. (2006). Alginate hydrogels as biomaterials. Macromol Biosci, 6, 623–633. doi: 10.1002/mabi.200600069.

[4] Beamish, J. A., Zhu, J., Kottke-Marchant, K., & Marchant, R. E. (2010). The effects of monoacrylated poly (ethylene glycol) on the properties of poly (ethylene glycol) diacrylate hydrogels used for tissue engineering. J Biomed Mater Res Part A, 92, 441–450. doi: 10.1002/jbm.a.32353.

[5] Hunkeler, D. (1992). Synthesis and characterization of high molecular weight water-soluble polymers. Polym Int, 27, 23–33.

[6] Buchholz, F. L. & Graham, A. T. (1998). Modern Superabsorbent Polymer Technology. New York, Wiley- VCH.

[7] Chen, J., Park, H., & Park, K. (1999). Synthesis of superporous hydrogels: Hydrogels with fast swelling and superabsorbent properties. J Biomed Mater Res, 44, 53–62.

[8] Brannon-Peppas, L. & Harland, R. S. (1991). Absorbent polymer technology. J Control Release, 17(3), 297–298.

[9] Zohuriaan-Mehr, M. J. & Kabiri, K. (2008). Superabsorbent polymer materials: A review. Iran Polym J, 17(6), 451–477.

[10] Jeong, K. H., Park, D., & Lee, Y. C. (2017). Polymer-based hydrogel scaffolds for skin tissue engineering applications: A mini-review. J Polym Res, 24, 112. https://doi.org/10.1007/s10965-017-1278-4.

[11] Lee, R. H., Kim, B., Choi, I., Kim, H., Choi, H. S., Suh, K., Bae, Y. C., Jung, J. S. (2004). Characterization and expression analysis of mesenchymal stem cells from human bone marrow and adipose tissue. Cell Physiol Biochem, 14(4–6), 311–324. doi: 10.1159/000080341. PMID: 15319535.

[12] Hou, L., Cao, H., Wang, D. et al. (2003). Induction of umbilical cord blood mesenchymal stem cells into neuron-like cells in vitro. Int J Hematol, 78, 256–261. https://doi.org/10.1007/BF02983804.

[13] Lanthong, P., Nuisin, R., & Kiatkamjornwong, S. (2006). Graft copolymerization characterization and degradation of cassava starch-g acrylamide/itaconic acid superabsorbents. Carbohydr Polym, 66, 229–245.

[14] Patel, D., Sharma, S., Screen, H. R., & Bryant, S. J. (2018). Effects of cell adhesion motif, fiber stiffness, and cyclic strain on tenocyte gene expression in a tendon mimetic fiber composite hydrogel. Biochem Biophys Res Commun, 499, 642–647. doi: 10.1016/j.bbrc.2018.03.203.

[15] Rodríguez-Rodríguez, R., García-Carvajal, Z., Jiménez-Palomar, I., Jiménez-Avalos, J., & Espinosa-Andrews, H. (2019). Development of gelatin/chitosan/PVA hydrogels: Thermal stability, water state, viscoelasticity, and cytotoxicity assays. J Appl Polym Sci, 136, 47149. doi: 10.1002/app.47149.

[16] Shi, J., Votruba, A. R., Farokhzad, O. C., & Langer, R. (2010). Nanotechnology in drug delivery and tissue engineering: From discovery to applications. Nano Lett, 10, 3223–3230. doi: 10.1021/nl102184c.

[17] Anisha, S., Kumar, S. P., Kumar, G. V., & Garima, G. (2010). Hydrogels, 4(2). Article 016. ISSN: 0976-044X [September–October].

[18] Slaughter, B. V., Khurshid, S. S., Fisher, O. Z., Khademhosseini, A., & Peppas, N. A. (2009). Hydrogels in regenerative medicine. Adv Mater, 21, 3307–3329. doi: 10.1002/adma.200802106.

[19] Sun, J.-Y., Zhao, X., Illeperuma, W. R., Chaudhuri, O., Oh, K. H., Mooney, D. J., et al. (2012). Highly stretchable and tough hydrogels. Nature, 489, 133–136. doi: 10.1038/nature11409.

[20] Qunyi, T. & Ganwei, Z. (2005). Rapid synthesis of a superabsorbent from a saponified starch and acrylonitrile/AMPS graft copolymers. Carbohydr Polym, 62, 74–79.

[21] Vasile, C., Pamfil, D., Stoleru, E., & Baican, M. (2020). New developments in medical applications of hybrid hydrogels containing natural polymers. Molecules, 25, 1539. doi: 10.3390/molecules25071539.

[22] Wang, C., Kopecek, J., & Stewart, R. J. (2001). Hybrid hydrogels cross-linked by genetically engineered coiled-coil block proteins. Biomacromolecules, 2, 912–920. doi: 10.1021/bm0155322.

[23] Ma, Y. & Lee, P. (2009). Investigation of suspension polymerization of hydrogel beads for drug delivery. Iran Polym J, 18(4), 307–313.

[24] Tabata, Y. (2009). Biomaterial technology for tissue engineering applications. J R Soc Interf, 6, S311–S324.

[25] Yang, G., Lin, H., Rothrauff, B. B., Yu, S., & Tuan, R. S. (2016). Multilayered polycaprolactone/gelatin fiber-hydrogel composite for tendon tissue engineering. Acta Biomater, 35, 68–76. doi: 10.1016/j.actbio.2016.03.004.

[26] Yang, J. M., Su, W. Y., & Yang, M. C. (2004). Evaluation of chitosan/PVA blended hydrogel membranes. J Memb Sci, 236, 39–51. doi: 10.1016/j.memsci.2004.02.005.

[27] Yang, J., Xu, C., Wang, C., & Kopecek, J. (2006). Refolding hydrogels self-assembled from *N*-(2-hydroxypropyl) methacrylamide graft copolymers by antiparallel coiled-coil formation. Biomacromolecules, 7, 1187–1195. doi: 10.1021/bm051002k.

[28] Li, Y., Huang, G., Zhang, X., Li, B., Chen, Y., Lu, T., Lu, T. J., & Xu, F. (2013). Magnetic hydrogels and their potential biomedical applications. Adv Funct Mater, 23(6), 660–672.

[29] Zhang, L., Gu, F., Chan, J., Wang, A., Langer, R., & Farokhzad, O. (2008). Nanoparticles in medicine: Therapeutic applications and developments. Clin Pharmacol Ther, 83, 761–769. doi: 10.1038/sj.clpt.6100400.

[30] Zhang, T., Yang, R., Yang, S., Guan, J., Zhang, D., Ma, Y., et al. (2018). Research progress of self-assembled nanogel and hybrid hydrogel systems based on pullulan derivatives. Drug Deliv, 25, 278–292. doi: 10.1080/10717544.2018.1425776.

[31] Zustiak, S. P., Pubill, S., Ribeiro, A., & Leach, J. B. (2013). Hydrolytically degradable poly (ethylene glycol) hydrogel scaffolds as a cell delivery vehicle: Characterization of PC12 cell response. Biotechnol Prog, 29, 1255–1264. doi: 10.1002/btpr.1761.

Shehla Khan, Abdur Rauf*, Zubair Ahmad, Hassan A. Hemeg

Chapter 5
Hydrogel in wound dressing and burn dressing products with antibacterial potential

Abstract: Infections in wounds and the development of sepsis pose significant risks for patients with burn injuries as thermal injury suppresses the immune system and makes the patients susceptible to life-threatening conditions. The most advanced wound treatment nowadays is the controlled delivery and the release of active substances into the burn sites. Hydrogels are, therefore, ideal for burn wounds due to their unique properties like high water content, nonadhesiveness to burn wounds, flexibility, and biocompatibility to provide a cooling environment and the ability to absorb extra exudate from the wounds. Various advanced hydrogels are nowadays available in the market, which have proven positive and effective results clinically against various burn wounds along with microbial infections. This chapter documents the advantages of hydrogels over traditional burn wound treatments and the applications of various hydrogels with antibacterial potential in burn wounds.

Keywords: Burn wounds, hydrogels, antibacterial potential, wound treatments

5.1 Introduction

A hydrogel is a three-dimensional (3D) network of hydrophilic polymers that can absorb and retain a substantial amount of water while preserving its structure through physical and chemical cross-linking of the polymer chains. The concept of hydrogels was initially introduced by Wichterle and Lím in 1960 [1, 2]. For a material to be classified as a hydrogel, it should contain a minimum of 10% water in relation to its total weight or volume. Hydrogels emulate the pliability of natural tissues, owing to their

*Corresponding author: Abdur Rauf, Department of Chemistry, University of Swabi, Khyber Pakhtunkhwa, Anbar 23561, Pakistan, e-mail: mashaljcs@yahoo.com
Shehla Khan, Department of Biotechnology, University of Swabi, Swabi, Anbar, Khyber Pakhtunkhwa, Pakistan
Zubair Ahmad, Department of Chemistry, University of Swabi, Khyber Pakhtunkhwa, Anbar 23561, Pakistan
Hassan A. Hemeg, Department of Medical Laboratory Technology, College of Applied Medical Sciences, Taibah University, P.O. Box 344, Al-Medinah Al-Monawara 41411, Saudi Arabia

https://doi.org/10.1515/9783111334080-005

high water content. A hydrogel is a cross-linked hydrophilic polymer that is not dissolved in water. They maintain a well-defined structure and are highly absorbent. These properties are related to various applications of hydrogels, specifically in the biomedical field. Most of the hydrogels are synthetic; however, some are also obtained from nature [3, 4].

5.2 Types of hydrogels

Hydrogels are classified into different classes on the basis of various factors such as source, origin, structural configuration, composition, cross-linking, durability, response to external stimuli, and network charge [5]. Here the classification on the basis of sources will be discussed. On the basis of source, there are three different classes of hydrogels, such as synthetic, natural, and hybrid hydrogels, as shown in Figure 5.1.

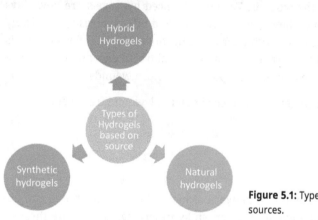

Figure 5.1: Types of hydrogels based on the sources.

5.2.1 Synthetic hydrogels

Polymeric synthetic hydrogels are 3D swelling structures of ionically or covalently cross-linked hydrophilic homopolymer or copolymer hydrogels. Synthetic hydrogels are developed by the polymerization of various synthetic monomers. Some of the synthetic hydrogels include poly(hydroxyethyl methacrylate), polyethylene glycol, and polyacrylic acid hydrogels [6].

5.2.2 Natural hydrogels

Natural hydrogels have strong cell adhesion properties. These hydrogels are biocompatible and biodegradable. Natural polymers such as proteins and polysaccharides are used to yield natural hydrogels. These proteins include gelatin, lysozyme, and collagen, whereas alginate and chitosan are the polysaccharides used for natural hydrogel production [7].

5.2.3 Hybrid hydrogels

Hybrid hydrogels are synthesized by the combination of synthetic and natural polymer hydrogels. Various natural biopolymers such as dextran, chitosan, and collagen are combined with synthetic polymers such as alcohol and poly-*N*-isopropyl-acrylamide. Examples of hybrid hydrogel are alginate/poly(polyethylene glycol) hydrogel and chitosan/polyvinyl alcohol hydrogel [8].

5.3 Applications of hydrogels

Hydrogels have a wide range of applications in various biomedical fields: bioseparation, electrophoresis, proteomics, tissue engineering chromatography, contact lenses, wound healing, and drug delivery. They are well-known in food and medicine, as filters in water purification, as absorbents in disposable diapers, and as separation materials for electrophoresis and chromatography. They are also of interest for controlled drug release and for the concentration of macromolecules in dilute solutions [9]. This chapter explains the importance and types of hydrogels used in burn wounds and other wound infections with potential antibacterial properties.

5.3.1 Hydrogels in burn wounds with antibacterial potential

One of the most important applications of hydrogels is in the burn wounds. Burns are defined as severe skin damage in which the affected cells of the skin die. When infection occurs in burn wounds, it sometimes leads to sepsis, disability, poor healing, and even can be fatal. Traditionally, care mostly focuses on the early removal of dead tissues from the wounds, resuscitation of fluid, and intravenous antibiotics, but these are often considered insufficient due to compromised vasculature that limits the efficacy of the antibiotics. The presence of biofilm in burn wounds acts as a barrier to various wound treatments. These biofilms also cause the transition of wounds from acute to chronic form and to a nonhealing state. Burn wounds are currently treated with topical treatments that include skin substitutes such as stem and skin cells, and silver, or synthetic antimicrobial peptides (AMPs) or hydrogels. The skin substitute assists in the healing process while the AMPs/hydrogels deliver an antibiotic to the wound's site [9]. Infections and sepsis in wounds are serious complications for burn patients as thermal injury induces a state of immune suppression that makes the patients susceptible to a serious condition. A high density of microbial population can cause biofilm growth on the graft bed or burn wound. Indeed, most burn wounds are colonized by bacteria forming biofilm, due to which the healing process is disrupted significantly which results in impaired structural and functional wound tissues [10]. Over the last two decades, the survival rate of burn patients has improved significantly; however, sepsis, inflammatory response syndrome, and multiple organ failure syndrome are still the major causes of morbidity and mortality for these patients [11]. The most advanced wound treatment nowadays is the controlled administration delivery and release of active products into the burn sites. Hydrogels for burn wounds have numerous advantages such as nonadhesiveness to burn wounds, their ability to extract extra exudates from wounds, and to provide a cooling and moist environment. Due to these advantages, they are the most ideal for wound dressings. Also, hydrogels due to their large water content mimic the physiological conditions of wounds, which favors the tissues to regenerate with excellent biocompatibility. The hydrogels also have the ability to encapsulate a large number of various antimicrobial drugs. This ability additionally provides protection from various pathogens, hence providing favorable conditions that can promote the natural self-healing process for rigorous healing [12]. Hydrogel dressings are therefore known as significant drug carriers for the next generation of products having therapeutic efficacy and improved sustainability. Moreover, hydrogel synthesis is known for its capability to calibrate their various biological, chemical, and physical properties according to the method of treatment [13]. The chemical and physical properties of hydrogels can be fine-tuned for the optimum release of the antimicrobial drugs encapsulated in them against various bacterial biofilms. Such optimum release, sustained release, and slow release of the drug have proven to reduce the toxicity to mammalian cells. The extent of cross-linking of macromolecules in the hydrogels allows the calibration of its physical, chemical, and biological properties [13]. Many natural hydrogels such as β-chitin, chitosan, cellulose, and dextran are naturally embedded with antimicrobial properties, while many of synthetic hydrogels are encapsulated with antibacterial products such as synthetic antibiotics. These loaded hydrogels include AMP-based hydrogel, metallic nanoparticles, metal ion hydrogel, and natural polymer hydrogels [14, 15]. The fabrication of novel antimicrobial hydrogels and stimulus-responsive drug delivery hydrogels has increased in recent years due to the progress in technology and medicine. These novel hydrogels that are capable of combating wound infections include photo-controlled release hydrogels, thermogels, multiresponsive hydrogels, and magnetic gels [16]. The status and progress of wound healing can also be monitored by utilizing novel designed techniques such as the inclusion of sensor molecules into the hydrogels [13].

Hydrogels are clinically used as moist dressings, components of wound treatments, and wound debridement agents. In the management of burn wounds, hydrogel works as a moisture donor, thus accelerating the healing of wounds through moisture regulation and autolytic debridement [17]. The testing of stimulus-responsive drug delivery hydrogels and novel antimicrobial hydrogels has been increased in recent years due to the generation of animal models with acute and biofilm wound infections [14, 18, 19].

5.3.2 EPL hydrogels

EPL (epsilon-poly-L-lysine) is a thermostable, safe, and an antimicrobial agent. It has been approved by the administrations of the USA, Korea, and Japan, and is widely used as a food preservative [9]. EPL has a broad-spectrum antibacterial property, that is, it is equally effective against both gram-positive bacteria and gram-negative bacteria [20]. Therefore, EPL-catechol hydrogel possesses both antimicrobial and antibiofilm efficacies against multidrug-resistant bacteria. The EPL-catechol hydrogels are synthesized in a chemical reaction via cross-linking in situ of catechol and amine groups of EPL molecules [21, 22]. During the chemical reaction, the catechol molecule transforms the color into brown due to the production of *o*-quinone by the oxidation reaction [23].

In developing countries, *Acinetobacter baumannii* infection is significantly increased in patients with burn wounds. In a hospital environment, this infection is viable for a long time and is easily transmitted due to resistance to desiccation, multidrug resistance, and its capability of adherence to inanimate surfaces. Khan et al. utilizing the partial-thickness model of burn wounds followed by infection of gram-negative *Acinetobacter baumannii* suggested that EPL hydrogels remarkably reduced the in vivo biofilms of *A. baumannii* in burn wounds. The reduction is most likely due to the action of EPL and the function of reactive oxygen species and H_2O_2 by-products of catechol oxidation [24]. This is in agreement with the previous studies in which injectable EPL hydrogels were used in vivo against infections and for tissue healing and regeneration [25]. The exact mechanisms of EPL action are, however, yet to be determined in burn wounds. The EPL-catechol biofunctions against infections in burn wounds making it a promising material for future research in the biomedical field. Using cell lines and in vivo tissue analysis, the EPL hydrogel is also believed to have no toxicity or inflammation upon its topical or intradermal applications [17].

5.3.3 Hydrogel in wound dressing with antibacterial potential

Any disruption or defect in the skin that can result from trauma or physiological/medical condition is known as a wound. Wound healing requires special care and special dressing material. Effective dressing design depends on knowing the healing process, the effect of each dressing material on the skin, and on specific patient conditions as well [26, 27]. Healing of wounds can be affected by numerous factors such as the unusual presence of bacteria or infection, necrosis, maceration, edema, and trauma [28]. Depending on the skin's affected area or damaged number of skin layers, wounds can be classified into three different types such as superficial, partial-thickness, and full-thickness wounds [26].

5.3.3.1 Superficial wounds

Superficial wounds are wounds in which only the epidermis layer of the skin is involved and damaged.

5.3.3.2 Partial-thickness wounds

These are the wounds in which the epidermis and deeper dermal layers are affected.

5.3.3.3 Full-thickness wounds

When deeper tissue of the skin and subcutaneous fat has been damaged, then that wound is classified as full-thickness wound.

5.4 Wound dressing

The absolute wound dressing material has many factors involved, which make it an ideal therapy for wound management. It should absorb excess toxins and exudates from the skin, keep a balanced moisture between the dressing and the wound, prevent the wound site from excess heat, that is, provide a cooling effect, safeguard the wound from any external infection, should be completely sterilized and easily removed from the wound site without further damaging the skin, and should have good permeability to gases [29]. The choice of an ideal dressing for a particular wound is important for optimum healing [1]. For this purpose, the wound dressing industry recently focused on the need to provide comfort wound dressing material, its long shelf life, and cost-effectiveness which together make it perfect for a specific patient [29].

5.4.1 Types of wound dressing materials

Most of the currently available wound dressing materials are categorized as semiper-meable films, low adherent wound dressing, hydrocolloids, alginates, antimicrobial dressing, foam dressing, and most importantly hydrogels. Plain gauze is still the most utilized dressing material for various types of wounds; however, new research and development in the field of wound management has developed advanced healing products with improved chemical and physical properties (Table 5.1) [30].

Table 5.1: Advanced wound dressings listed from Murphy [30].

Class	Type	Details
Protective dressing	Gauze	Readily accessible and cost-effective
	Impregnated gauze	Nonadherent, retains moisture
Antimicrobial dressings	Antibacterial ointments	Often reapplied to maintain moisture
	Iodine-based	Unsuitable for individuals with thyroid disorders
	Silver-based	Available in various forms; offers broad-spectrum efficacy with low resistance
Autolytic debridement	Films	Occlusive; allows for gas exchange
	Hydrocolloids	Not recommended for exudative or infected wounds
	Hydrogels	Rehydrates to soften dry wounds
Chemical debridement	Papain/urea	Availability concerns in the USA
	Collagenase	Facilitates selective debridement
	Foam	Absorbs moderate exudate
	Absorbs moderate exudate	Suitable for minimal exudate
	Hydrofibers	Effective for heavy exudate
	Alginates	Ideal for absorbing heavy exudate

Advanced wound dressing products are designed to keep a balance of moisture at the wound's site, which allows the fluid to remain intact to the wounds without spreading to healthy unaffected skin tissues [31]. It was first discovered by winter in 1962 that a moist environment at the wound site has a substantial accelerating effect on the heal-ing process of the wounds [32]. Hydrocolloids and hydrogels are the dressing materi-als designed for moist wound healing. Both of these dressing products promote the removal of dead tissue by inducing autolytic debridement at the wound site [28]. Hydrocolloids are mainly composed of gelatin, sodium carboxymethylcellulose, adhesives, elastomers, and pectin.

Hydrofiber wound dressings (ConvaTec) are another form of ideal dressing materials. These materials form a swollen gel structure at the wound's site by seizing the moisture, thus accelerating the healing process. Hydrofiber dressings are in the form of nonwoven hydrophilic flat sheets which upon absorbing the wound exudates are converted into soft gel sheets [29].

5.4.2 Hydrogels in wound dressing

Hydrogels, another form of wound dressings, are widely used for the removal of damaged tissues or foreign objects from (a wound) making it one of the best debriding agents. It retained and absorbed the exudate from the contaminated site within the gel mass. This is done by cross-linking polymer chains in the hydrogels, which results in the isolation of debris, odor molecules, and bacteria in the liquid. The high water content of hydrogels makes it an ideal dressing for various wound sites such as leg ulcers, pressure sores, necrotic/surgical wounds, burns, and lacerations, by allowing oxygen and vapor transmission to these sites. Due to their hydrating and cooling effect, they play a vital role in emergency burn treatments alone or in combination with other dressing materials [32]. One such example is Burnshield hydrogel (Levtrade International), which is available in first aid kits as emergency burn dressing. It is present in the form of polyurethane foam and is composed of 1.06% *Melaleuca alternifolia* extract and 96% of water [2]. Sheet hydrogels are mostly used and changed two to three times a week while amorphous hydrogels are usually applied daily on the wounds [28]. Hydrogel wound dressings are also utilized for granulating wound cavities [33, 34] (Table 5.2).

Table 5.2: A few hydrogels and hydrogel sheets used as wound dressings.

Dressing materials	Components	Characteristics
Granugel[R] (ConvaTec)	Pectin, carboxymethylcellulose, and propylene glycol	A clear, viscous hydrogel designed for managing partial- and full-thickness wounds. It can also serve as a filler for dry cavity wounds, creating a moist healing environment.
Intrasite Gel[R] (Smith & Nephew)	Modified carboxymethylcellulose (2.3%) and propylene glycol (20%)	This amorphous sterile hydrogel dressing is suitable for both shallow and deep open wounds.
Purilon Gel[R] (Coloplast)	Sodium carboxymethylcellulose and more than 90% of water	Indicated for use in conjunction with a secondary dressing for necrotic and sloughy wounds, as well as first and second-degree burns

Table 5.2 (continued)

Dressing materials	Components	Characteristics
Aquaflo™ (Covidien)	Polyethylene glycol/propylene glycol	Designed in a disk shape to maximize wound coverage and aid in filling shallow cavities. The translucent gel allows for easy wound visualization.
Woundtab[R] (First Water)	Sulfonated copolymer, carboxymethylcellulose, glycerol, and water	This dressing contains a superabsorbent polymeric gel that can absorb bacteria and retain them within its structure. It is described as a wound "kick-starter" patch for chronic wounds and can also be utilized as a secondary absorbent.

5.4.3 Hydrogel with antibacterial potential

Cartmell and Sturtevant in 1992 developed a wound dressing that is transparent and thin film with a central nonadhesive area. This dressing contains hydrogel that is composed of isophorone diisocyanate and polyethylene glycol or polypropylene glycol. This specific hydrogel is proposed to be flexible as its removal is easy, and due to its transparent nature, the wound healing process can be constantly observed. The edges of these hydrogel dressings stick to the surface of the skin due to their adhesive layer which in turn safeguards the wounds from any foreign bodies and bacteria [35]. If there is any systemic or local infection that is hindering the wound healing process, in that case, one possible therapeutic perspective is to use hydrogel dressing that possesses various antimicrobial products to combat infections such as silver or iodine. Silver is a natural antibacterial agent effective against various microbes including *Staphylococcus aureus* and *Pseudomonas aeruginosa* [36]. These pathogens are opportunistic microbes and are usually found in chronic wounds. The mode of action of these microbes is biofilm production in host cells [37]. In a biofilm, critical colonization occurs due to bacterial multiplication which is usually accompanied by an increased pain at the infection site. This critical colonization can hinder or delay the healing process due to the development of a thick slough that is nonresponsive to standard dead and damaged tissue removal techniques. To combat this situation, it is important to minimize the bacterial levels to a minimum in order to accelerate the wound healing process. This can only be achieved by using topical antimicrobial dressing to lower the bacterial population [36]. In this regard, the US patent 8,431,151 B2 manufactured an antimicrobial hydrogel dressing with a controlled release of silver ions. They directly incorporated the silver nitrate ions into polymer/dimethyl form amide solution to produce the antibacterial scaffold [38].

Continued demands from healthcare professionals and the public for wound dressing products can lead to future and advanced developments in this field [39]. An important future challenge is to produce an appropriate wound healing strategy for every patient by providing optimal dressing products. There is a need to manufacture innovative dressing while keeping the production cost low.

5.5 Conclusion

The management of burn injuries, characterized by an increased susceptibility to infections and sepsis due to compromised immunity, has witnessed a transformative approach through the controlled delivery and release of active substances at burn sites. Hydrogels have emerged as an ideal solution for burn wounds, owing to their unique properties such as high water content, nonadhesiveness to burn wounds, flexibility, biocompatibility, and their capacity to provide a cooling environment while absorbing excess exudates from wounds. The market today offers a wide array of advanced hydrogel products that have demonstrated positive and effective clinical outcomes in the treatment of various burn wounds, including those complicated by microbial infections. These hydrogels, with their antibacterial potential, represent a promising advancement in burn wound management, highlighting the advantages of hydrogels over traditional burn wound treatments and their diverse applications in improving patient outcomes while minimizing the risks associated with burn-related infections, ultimately enhancing the overall quality of care for burn injury patients.

References

[1]   Beldon, P. (2010). How to choose the appropriate dressing for each wound type. Wound Essen, 5, 140–144.
[2]   Burnshield Emergency Burncare website. (2014). https://www.burnshield.com
[3]   Ahmed, E. M. (2015). Hydrogel: Preparation, characterization, and applications: A review. J Adv Res, 6(2), 105–121.
[4]   Cai, W. & Gupta, R. B. (2012). Hydrogels. Kirk-Othmer Encycl Chem Technol. doi: 10.1002/0471238961.
[5]   Bashir, S., Hina, M., Iqbal, J., Rajpar, A. H., Mujtaba, M. A., Alghamdi, N. A., Wageh, S., Ramesh, K., & Ramesh, S. (2020). Fundamental concepts of hydrogels: Synthesis, properties, and their applications. Polymers (Basel), 12(11), 2702.
[6]   Vijayavenkataraman, S., Vialli, N., Fuh, J. Y., & Lu, W. F. (2019). Conductive collagen/polypyrrole-b-polycaprolactone hydrogel for bioprinting of neural tissue constructs. Int J Bioprinting, 5, 229.
[7]   Singh, S. K., Dhyani, A., & Juyal, D. (2017). Hydrogel: Preparation, characterization and applications. Pharma Innov, 6, 25.
[8]   Zaman, M., Siddique, W., Waheed, S., Sarfraz, R., Mahmood, A., Qureshi, J., Iqbal, J., Chughtai, F. S., Rahman, M. S. U., & Khalid, U. (2015). Hydrogels, their applications and polymers used for hydrogels: A review. Int J Biol Pharm Allied Sci, 4, 6581–6603.

[9]   Yamanaka, K. & Hamano, Y. (2010). Biotechnological production of poly-epsilon-l-lysine for food and medical applications. In: Hamano, Y. (ed.), Amino-Acid Homopolymers Occurring in Nature. Springer, 61–75.

[10]   Mahmoudi, H., Pourhajibagher, M., Chiniforush, N., Reza-Soltanian, A., Alikhani, M. Y., & Bahador, A. (2019). Biofilm formation and antibiotic resistance in methicillin-resistant and methicillin-sensitive Staphylococcus aureus isolated from burns. J Wound Care, 28, 66–73.

[11]   Norbury, W., Herndon, D. N., Tanksley, J., Jeschke, M. G., & Finnerty, C. C. (2016). Infection in burns. Surg Infect, 17(2), 250–255.

[12]   Wang, Y., Beekman, J., Hew, J., Jackson, S., Issler-Fisher, A. C., Parungao, R., Lajevardi, S. S., Li, Z., & Maitz, P. K. M. (2018). Burn injury: Challenges and advances in burn wound healing, infection, pain and scarring. Adv Drug Deliv Rev, 123, 3–17.

[13]   Francesko, A., Petkova, P., & Tzanov, T. (2018). Hydrogel dressings for advanced wound management. Curr Med Chem, 25, 5782–5797.

[14]   Haidari, H., Kopecki, Z., Bright, R., Cowin, A. J., Garg, S., Goswami, N., & Vasilev, K. (2020). Ultrasmall AgNP-impregnated biocompatible hydrogel with highly effective biofilm elimination properties. ACS Appl Mater Interfaces, 12, 41011–41025.

[15]   Zhong, Y., Xiao, H., Seidi, F., & Jin, Y. (2020). Natural polymer-based antimicrobial hydrogels without synthetic antibiotics as wound dressings. Biomacromolecules, 21, 2983–3006.

[16]   Trombino, S., Servidio, C., Curcio, F., & Cassano, R. (2019). Strategies for hyaluronic acid-based hydrogel design in drug delivery. Pharmaceutics, 11(8), 407.

[17]   Zhang, W., Wang, R., Sun, M. S., Zhu, B. X., Zhao, Q., Zhang, T., Cholewinski, A., Yang, F. M., Zhao, B., Pinnaratip, R., Forooshanie, P. K., & Bruce, B. P. (2020). Catechol-functionalized hydrogels: Biomimetic design, adhesion mechanism, and biomedical applications. Chem Soc Rev, 49, 433–464.

[18]   Kopecki, Z. & Cowin, A. (2017). Fighting chronic wound infection: One model at a time. Wound Pract Res, 25, 6–13.

[19]   Ogunniyi, A. D., Kopecki, Z., Hickey, E. E., Khazandi, M., Peel, E., et al. . (2018). Bioluminescent murine models of bacterial sepsis and scald wound infections for antimicrobial efficacy testing. PLOS ONE, 13(7).

[20]   Zhou, C., Li, P., Qi, X., Sharif, A. R. M., Poon, Y. F., & Cao, Y., et al. . (2011). A photopolymerized antimicrobial hydrogel coating derived from epsilon-poly-l-lysine. Biomaterials, 32(2704–2712).

[21]   Byun, E. & Lee, H. (2014). Enhanced loading efficiency and sustained release of doxorubicin from hyaluronic acid/graphene oxide composite hydrogels by a mussel-inspired catecholamine. J Nanosci Nanotechnol, 14, 7395–7401.

[22]   Qiu, W. Z., Wu, G.-P., & Xu, Z. K. (2018). Robust coatings via catechol-amine codeposition: Mechanism, kinetics, and application. ACS Appl Mater Interfaces, 10, 5902–5908.

[23]   Hou, J., Li, C., Guan, Y., Zhang, Y., & Zhu, X. X. (2015). Enzymatically crosslinked alginate hydrogels with improved adhesion properties. Polym Chem, 6, 2204–2213.

[24]   Khan, A., Xu, M., Wang, T., You, C., Wang, X., Ren, H., Zhou, H., Khan, A., Han, C., & Li, P. (2019). Catechol cross-linked antimicrobial peptide hydrogels prevent multidrug-resistant Acinetobacter baumannii infection in burn wounds. Biosci Rep, 39(6), 18.

[25]   Sun, A., He, X., Li, L., et al. (2020). An injectable photopolymerized hydrogel with antimicrobial and biocompatible properties for infected skin regeneration. NPG Asia Mater, 12(25).

[26]   Eccleston, G. M. (2007). The Design and Manufacture of Medicines (3rd ed., pp. 598–605). Aulton's Pharmaceutics, Churchill Livingston Elsevier.

[27]   Medaghiele, M., Demitri, C., Sannino, A., & Ambrosio, L. (2014). Polymeric hydrogels for burn wound care: Advanced skin wound dressings and regenerative templates. Burns Trauma, 2(4), 153–161.

[28]   Turner, T. D. (1979). Hospital usage of absorbent dressings. Pharm J, 222, 421–424.

[29]   Jones, V., Grey, J. E., & Harding, K. G. (2006). Wound dressings. Br Med J, 332, 777–780.

[30]   Murphy, P. S. & Evans, G. R. D. (2012). Plast Surg Int, 2012, 190436.

[31] Stashak, T. S., Farstvedt, E., & Othic, A. (2004). Update on wound dressings: Indications and best use. Clin Tech Equine Pract, 2004(3), 148–163.

[32] Osti, E. & Osti, F. (2004). Treatment of cutaneous burns with Burnshield (hydrogel) and a semi-permeable adhesive film. Ann Burn Fire Disasters, 3, 137–141.

[33] Wichterle, O. & D. Lím, D. (1960). Hydrophilic gels for biological use. Nature, 185, 117–118.

[34] Jones, A. & Vaughan, D. (2005). Hydrogel dressings in the management of a variety of wound types: A review. J Orthopaed Nurs, 9(1), 1–11.

[35] Cartmell, J. V. & Sturtevant, W. R. (1992). US Patent, 5,106,629.

[36] Flores, A. & Kingsley, A. (2007). Topical antimicrobial dressings: An overview. Wound Essen, 2, 182–185.

[37] Fazli, M., Bjarnsholt, T., Kirketerp-Møller, K., Jørgensen, B., Andersen, A. S., Krogfelt, K. A., Givskov, M., & Tolker-Nielsen, T. (2009). Nonrandom distribution of Pseudomonas aeruginosa and Staphylococcus aureus in chronic wounds. J Clin Microbiol, 47, 4084–4089.

[38] Mather, P., Wu, J., Ren, D., & Hou, S. (2013). US Patent 8,431,151 B2.

[39] Harding, K. G., Morris, H. L., & Patel, G. K. (2002). Science, medicine and the future: Healing chronic wounds. Br Med J, 324(7330), 160–163.

Farzana Nazir*, Khadija Munawar

Chapter 6
Emerging fabrication strategies of hydrogel and its use as a drug delivery vehicle

Abstract: Hydrogels are three-dimensional networks that have been recently investigated for their drug delivery applications. Hydrogels have the innate property of controlled drug release and quick response. Drug delivery of therapeutics, cells, and biomolecules involves a transporting vehicle such as hydrogels. Hydrogels have optimized drug loading and release. The first part focuses on various conventional and advanced strategies for the drug delivery of hydrogels such as the solvent casting method. Particulate leaching, freeze-drying, microfluidics, electrospinning, and lithography have been discussed in this chapter. The second part describes the various routes such as intramuscular (injection), transdermal, nasal, buccal, oral, and rectal drug delivery to polymer-based hydrogels for drug delivery applications.

Keywords: Hydrogels, Drug delivery, Methods of drug delivery, Polymer based hydrogels, Drug delivery routes

6.1 Introduction

Hydrogels have evolved into perfect materials for nascent applications like drug delivery and tissue engineering. Hydrogels are soft materials characterized by a polymer network surrounded by a solvent. In hydrogels, the polymer network has covalent bonds, coordinate covalent bonds, electrostatic interactions, and hydrophobic interactions [1].

Hydrogels have attracted the attention of biomaterial scientists due to their hydrophilic and biocompatible properties since the pioneering research by Wichterle and Lim in 1960 [2, 3]. Groundbreaking and significant contributions by Lim and Sun paved the way for cell encapsulation to treat diabetes by using calcium alginate in the 1980s [4]. In 1989, the natural polymer collagen-based hydrogel acting as an extracellular matrix for dermal and epidermal cells was developed for skin burn dressing [5]. Natural and synthetic polymers have been attractive candidates for hydrogel preparation in the tissue engineering field. Hydrogels mimic the matrices for repair and regeneration in tissue engineering.

*Corresponding author: Farzana Nazir, Department of Chemistry, School of Natural Sciences, National University of Sciences and Technology (NUST), Islamabad 44000, Pakistan, e-mails: farzananazir88@gmail.com, farzana.nazir@sns.nust.edu.pk
Khadija Munawar, Department of Chemistry, School of Natural Sciences, National University of Sciences and 3044 Technology (NUST), Islamabad (44000), Pakistan

https://doi.org/10.1515/9783111334080-006

The primary driving force behind tissue engineers' efforts over the past 20 years has been to offer material-based therapeutic solutions to increase the speed and quality of tissue defect repair or regeneration. The current paradigm involves co-delivering signaling chemicals like peptides and proteins, planted or encapsulated cells, and appropriate scaffolding materials. Almost all medication delivery approaches in tissue engineering require co-delivery of whole cells or the targeted release delivery of bioactive substances of various sizes [6].

Hydrogels are hydrophilic polymeric structures with water absorption limits ranging roughly from 10% to 20% to thousand times as compared to their dry weight. Hydrogels are classified as physical gels as well as chemical gels. Physical gels or reversible gel network is characterized by molecular entanglements and/or by secondary forces including ionic, H-bonding, or hydrophobic forces [7]. Calcium alginate is an example of this type of hydrogel. Chemical gels or permanent gel network is characterized by cross-linking. Gelatin methacrylate (GelMA) and hydroxyethyl methacrylate (HEMA) are examples of chemical gels [8].

Hydrogels have evolved into perfect materials for nascent applications like drug delivery and tissue engineering. Hydrogels are soft materials characterized by a polymer network surrounded by a solvent. In hydrogels, the polymer network has covalent bonds, coordinate covalent bonds, electrostatic interactions, and hydrophobic interactions. The abundance of hydrophilic groups, such as the hydroxyl and carboxyl groups, is responsible for water absorption and swelling even though hydrogels are resistant to dissolution in solvents [1]. Hydrogels do not dissolve in the solvent because of the cross-linking between the polymeric chains. Natural and synthetic polymers are used for the hydrogel preparation, and the selection of the polymer is defined by the hydrogel application and the site of drug delivery. The main feature of hydrogel is porosity and swelling, which can be modified by controlling cross-linking of the hydrogel. A porous network helps in micro- and macromolecules for drug loading and subsequent release by diffusion.

Hydrogels can be classified by size into three many categories: macroscopic hydrogels, microgels, and nanogels. Other types of hydrogels are core–shell structure, tetra-arm, topological, emulsion, multilayer, porous, and hybrid hydrogels. Hybrid hydrogels are classified into nanocomposites, macromolecular microsphere composites, and interpenetrating networks as shown in Figure 6.1 [9].

6.2 Hydrogel fabrication strategies for drug delivery

The goal of creating a drug delivery system is to make it possible to control drug release to treat medical diseases at a set pace for a certain amount of time. There are various advanced fabrication techniques for the development of hydrogels. These are physical techniques as well as chemical modifications to obtain hydrogels for drug delivery. Some of the fabrication techniques are listed in Figure 6.2.

Figure 6.1: Classification of hydrogels.

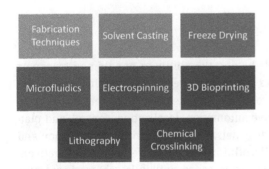

Figure 6.2: Fabrication techniques.

6.2.1 Physical methods of hydrogel preparation

Physical methods of hydrogel preparation are listed as solvent casting, freeze-drying, microfluidics, electrospinning, lithography and three-dimensional (3D) printing, emulsion, and self-assembling systems.

6.2.1.1 Solvent casting method: particulate leaching

Solvent casting aids in the regulated porosity or pore size. After hydration, solvent casting hydrogels swell and reveal a porous structure. It helps in the controlled porosity or the pore size solvent casting hydrogels swell after hydration and shows porous structure after water absorption [10, 11]. Disadvantages include the possibility of the solvent being retained in the scaffold if the solvent is not fully evaporated or in case of particulate leaching, the leaching agent may not be removed properly. Hydrogels resulting from this process have good isotropic properties [12, 13].

6.2.1.2 Freeze-drying

Freeze-drying is a very attractive technique that helps in the complete removal of the solvent from the hydrogel along with the formation of the highly connected porous scaffold. Freeze drying produces very small pores, however the porous structure compromises mechanical integrity [14]. Site-specific drug delivery can be efficiently obtained by using the freeze-drying method [15]. Texture and fast response of drug loading can be achieved by using this technique.

6.2.1.3 Microfluidics

Microfluidic devices are microchannels having the ability to handle and transport small fluid volumes. Microfluidic devices can be prepared by lithography for drug delivery vehicles. Efficiency and well control of the drug delivery to the target tissue can be done by microfluidics. Microfluids give automation, localization, control, and platform integration. Microfluids fabricate drug delivery systems with high precision and essential part of the cellular assay. Microfluidic microsphere technology delivers drugs such as proteins or peptides, nucleic acids, genes, antibiotics, chemotherapeutic drugs, growth factors, stem cells, and differentiated cells to the target tissue [16].

Microfluid technology gives high control of the different fluids to prepare microsphere hydrogels. Different devices used for microfluidic hydrogel microsphere are capillary microfluidic devices, microfluidic chips, in-air microfluidics, microfluidic chip fusion capillary device, and open-channel microfluidic device as shown in Figure 6.3.

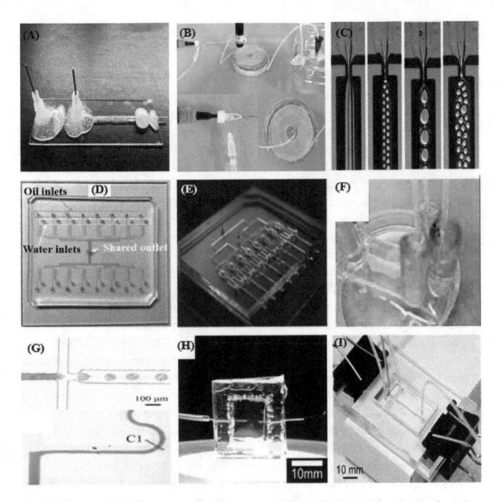

Figure 6.3: (A–C) Different capillary microfluidic devices; (D–H) microfluidic chips; and (I) open-channel microfluidic device [17] (Copyright 2021, reproduced with permission from the Royal Society of Chemistry).

Fluid dynamics such as the diameter, fluid displacement, and internal structure of the microfluid device help in optimizing the microfluidic hydrogel microspheres. Delivery systems based on technology have been widely used to deliver drugs in tissue engineering, including growth factors, proteins, genes, ions, and other drugs. Microfluidic devices hold the drug by covalent binding, conjugation, liposome, solution in water, and swelling and absorption [17].

6.2.1.4 Electrospinning

Electrospun nanofibers have a high surface-to-volume ratio, enhanced mechanical strength, and interconnected pore structure which ensure the rapid therapeutic uptake and discharge of drugs. A high-voltage source is applied to make micro- and nanofibers by using a polymer solution. Hydrogel swelling by volume change alters by changing the morphology of the material using electrospinning [18]. Genes, proteins, and enzymes, as well as a variety of medicines and biomolecules, have all been effectively inserted into electrospun nanofibers because of biocompatibility, mostly using two techniques: mixing electrospinning and coaxial electrospinning [19]. Electrospinning is also characterized by the method used such as divided into multineedle/multijet, needleless/needle free, multihole, free surface, blow assisted, colloid, emulsion, coaxial, side by side, and melt spinning. Electrospinning can be controlled by blending, fiber orientation, and targeted collectors [9].

Electrospinning is also a technique involving a sophisticated instrument having continuous processing and forming a highly interconnected structure. Electrospun fibers have a high surface-to-volume ratio with random orientation. The productivity of this technique is low, and limited scaffold thickness can be obtained [20, 21].

Drug and biological therapeutics are mixed with an electrospinning solution leading to drug loading in combination with hydrogels leading to controlled and sustainable drug release [22]. As compared to the blended systems, the drug release from electrospun fibers is slow and prolonged [23]. The challenges for drug delivery focus on the initial drug release and multiple drug-release profile without interfering with each other. Electrospun composites have emerged as a plausible solution for different drugs such as micro- and nanomaterials to be easily loaded with hydrogels [24]. For example, benzoin was encapsulated in poly-L-lactic acid (PLA) and BSA in chitosan (CS). The electrospun microsphere helped in studying the release of two drugs [25]. Indomethacin and doxorubicin were loaded on silica encapsulated by poly-ε-caprolactone and gelatin resulting in rapid release of indomethacin and sustained release of doxorubicin. In Figure 6.4, the SEM and fluorescent images show the silica core-shelled microsphere in electrospun fibers.

6.2.1.5 Lithography

Photolithography and soft lithography are two kinds of lithography. Photolithography transfers a photomask pattern by light exposure [27]. Soft lithography is a complementary version of photolithography, expanding the scope of lithography. A variety of elastomers such as polydimethylsiloxane (PDMS) is used in lithography. Soft lithography has the advantage of low-cost, simple, chemically inert, and biocompatible surface chemistry.

Figure 6.4: (a) SEM of the image of PCL/gelatin/silica nanoparticle electrospun fibers; (b) TEM image; and (c) fluorescent image showing silica microsphere fluorescence in yellow-green color. PCL/gelatin/silica nanoparticle composite [26] (Copyright (2013), reprinted with permission from the American Chemical Society).

6.2.1.6 Three-dimensional printing

Hydrogel synthesis by 3D printing is the most advanced technique that involves sophisticated instruments and multistep processes. 3D printing is an expensive technique with controlled quality and pore structure. Limitations of this method are filament resolution [28]. 3D printing involves handling low fluid volume, multiple and specific drug releases, and innovative composition and technologies. A growing field of interest in biomedical engineering and drug delivery is 3D printing technology, which may create complicated drug release patterns, precise drug dosage, new drug delivery systems, and 3D-printed polypills. 3D printing is an automated, personalized, precise, digitally controlled system, to make a layer-by-layer material deposition leading to 3D construction. First, 3D-printed Food and Drug Authority (FDA) approved the drug Spritam®, which led to the development of a drug delivery system [29]. The most crucial parameter for 3D printing is the development of a bio-ink with suitable drug loading, release, and bioadhesion. Both natural and synthetic polymers are used for 3D printing, such as collagen, gelatin, fibrin, silk, polyethylene glycol (PEG), alginate, cellulose, poly-L-lactic acid, poly-ε-caprolactone, and poly(lactic-co-glycolic) acid [30]. 3D printing has the advantage of multiple drug loadings in a single system and fine-tuning drug release. Drug delivery systems can be created using a variety of 3D printing processes. The most popular 3D printing technologies used in drug delivery include digital light processing, selective laser sintering, continuous liquid interface production, stereolithography, fused deposition modeling, material jetting, inkjet deposition, and binder jetting.

6.2.1.7 Cross-linking method

Various chemical modifications of polymers to obtain hydrogels for drug delivery are found in the literature. One of these is the cross-linking of polymers. In cross-linking we have three basic components: monomer, initiator, and cross-linker. The ratio of these three must not disturb the structural integrity of the hydrogel. Various natural and synthetic polymers are polyesters, carbohydrates, proteins, and lipids. The co-polymers used are made up of two monomers such as poly(acrylamide)-graft-pullulan copolymer. Some of the monomers for copolymers are listed in Figure 6.5. Cross-linking can be achieved by both physical and chemical methods, and some of these are discussed below. The viscoelasticity and mechanical properties of hydrogels for drug delivery can be achieved by this method.

Physical cross-linking

Physical cross-linking is the simplest form of cross-linking. Major phenomena inducing physical cross-linking are heating/cooling a polymer solution, ionic interaction, and h-bonding, by protein interaction. For example, gelatin and nanosilicate composites have electrostatic interaction resulting in shear-thinning hydrogels [31, 32].

Chemical cross-linking

Chemically cross-linked hydrogels show enhanced mechanical strength as compared to the starting polymers. Because of this interesting feature, chemically cross-linked hydrogels have emerged as an interesting area of research. Chemically cross-linked hydrogels are usually synthesized by different methods, such as chemical cross-linkers, addition reactions, aldehyde additions, free radical polymerization, and enzymatic and radiation cross-linking. Some of these are given below.

Functional groups such as NH_2, COOH, and OH have the innate property of hydrophilicity. Moreover, these functional groups may undergo chemical bonding such as Schiff base formation [33], amine carboxylic acid interaction [34], and an isocyanate. Addition reactions are another method of cross-linking, For instance, polysaccharides

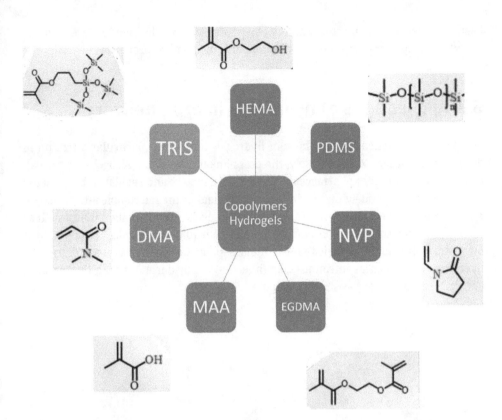

Figure 6.5: Copolymer hydrogels.

can be cross-linked by using 1,6-hexane dibromide [35]. Free radical polymerization is another method to develop hydrogels [36, 37]. Polysaccharides or polymers containing OH groups or proteins containing NH_2 can be cross-linked by using aldehyde; for example, polyvinyl alcohol (PVA) [38] and CS [39] can be linked through glutaraldehyde. The most advanced and fine-tunable hydrogels which can be synthesized by a 3D bioprinter are photo-cross-linkable hydrogels [40]. GelMA is a widely used photo-cross-linkable hydrogel for drug delivery, and recent advancements have led to the injectable GelMA for drug delivery [41, 42].

6.2.1.8 Blending method

Blends are another strategy to enhance biocompatibility and mechanical properties. Blending is much more cost-effective than chemical synthesis and thus it is a more frequently used method. The synergistic combination of two polymers in the blend affects physical properties such as glass transition temperature T_g, melting temperature T_m, crystallinity, and morphology as well as mechanical properties, degradation, barrier properties, and biocompatibility [43]. Hydrogels used by blending are a widely accepted method for drug delivery with desirable mechanical properties [44, 45].

6.3 Applications of hydrogels in drug delivery

The pragmatic physicochemical facet of hydrogels has drawn particular attention in drug administration. More precisely, the cross-linkable porous hydrogel structure allows for the coalition of pharmaceuticals in the gel matrix and regulates drug release at a rate that is dependent on the diffusion coefficient of the micromolecule or macromolecule across the gel network. In addition, hydrogels' exceptional qualities, such as their high water content, good biodegradability, adequate mechanical strength, high biocompatibility, and low toxicity make them popular for use in a variety of drug delivery methods, such as intramuscular (injection), transdermal, nasal, buccal, oral, and rectal drug delivery as shown in Figure 6.6.

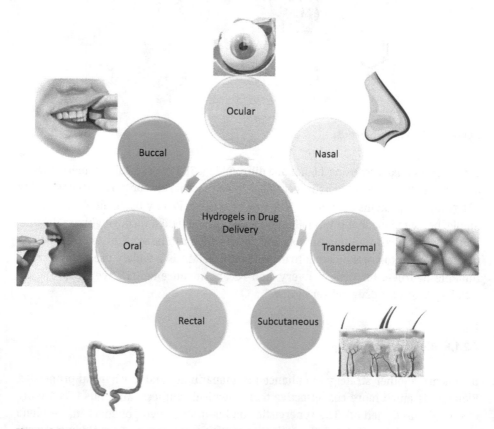

Figure 6.6: The application of hydrogels in different routes of drug delivery.

6.3.1 Ocular drug delivery

Eyes are the most sophisticated organ of the human body. The human eye consists of two major components: the anterior part is made of lens, cornea, pupil, iris, and ciliary muscle, and the posterior vitreous, sclera, retina, fovea, choroid, and optic nerve [46]. The human eye's function is impaired by visual diseases such as diabetic retinopathy, conjunctivitis, and retinal vascular disease [47].

Eye drops are the most common method of drug delivery for eye diseases. About 95% of the drug is lost in tears while only 5% of the drug reaches inside the eye tissue. Drug administered by eye drops in the eye is replenished with a constant flow of tear leading to very short contact time with eye tissue [48]. Consequently, the development of drug delivery involves major challenges such as increasing the drug residence time, prevention of adverse effects, and loss of drug. The eye drop is ineffective in glaucoma and dry eye treatments for chronic diseases. Many researchers have contemplated contact lenses as ocular drug delivery with the advantage of high efficacy, increased drug contact time, improved compliance, and bioavailability. Smart contact lenses structured from hydrogels have emerged as an ocular drug delivery approach because of transparency and biocompatibility.

Well-known materials for contact lenses are poly(methylmethacrylate), poly (ethylene terephthalate), poly(2-hydroxyethyl methacrylate), PDMS, silicone, and 2-methacryloyloxyethyl phosphorylcholine. The typical monomers used for hydrogels are dimethylacetamide, p(ethylene glycol dimethyl acrylate), methacrylic acid (MAA), 3-[tris(trimethylsiloxy)silyl]propyl methacrylate (TRIS), *N*-vinylpyrrolidone, and HEMA.

Researchers have reported the development of a novel, pH-sensitive hydrogel for potential use in ocular drug delivery by using p(HEMA-co-DPA) that was built on 2-(diisopropylamine) ethyl methacrylate (DPA) and 2-hydroxyethyl methacrylate (HEMA) as shown in Figure 6.7.

Hydrogel has the ability for controlling the release of dexamethasone 21-disodium phosphate as shown in Figure 6.8 for ophthalmological treatments at 6.5 pH. Hydrogel drug release efficacy was 70% during 20–40 h [49].

Levofloxacin, ofloxacin, moxifloxacin, and ibuprofen have been for drug delivery released by using thermosensitive hydrogel Pluronic F-127 [50–52]. Drug release was from 3 to 12 h with 80% to complete release.

CS-based hydrogels have been used to deliver latanoprost, ferulic acid, dexamethasone, and brimonidine tartrate (BT) using thermosensitive hydrogel for ocular drug delivery [53–56]. CS films prepared by dissolving in ionic liquid were solvent exchanged with plain water at room temperature. The resulting transparent, structurally stable, and mucoadhesive solvent-casted films delivered BT have high corneal permeability with fast drug release kinetics for potential ocular drug delivery [57]. These hydrogels have better results as compared to eye drops.

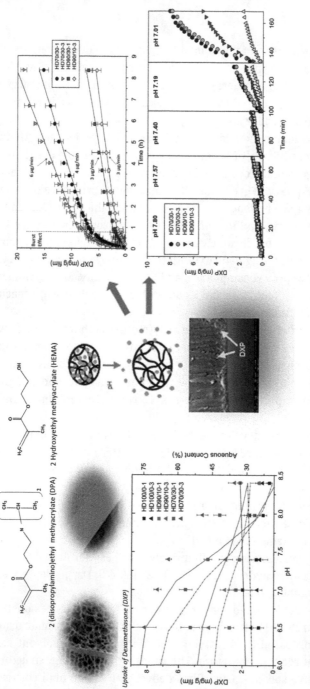

Figure 6.7: p(HEMA-co-DPA) built on 2-(diisopropylamine)ethyl methacrylate (DPA) and 2-hydroxyethyl methacrylate (HEMA) (reproduced after the permission of Elsevier, Copyright Faccia et al. [49]).

6.3.2 Nasal drug delivery

The nasal route of drug delivery is one of the first effects which increase drug bioavailability. Nasal mucosa has high vascularity, thinness, and substantial absorptive surface area, which helps drugs to enter systematic circulation and the nervous system. Nasal administration helps to surpass the blood–brain barrier (BBB) and avoids hepatic first-pass metabolism. Traditional drug delivery systems for the nasal route are nasal drops and sprays [58]. However, major limitations are absorption of macromolecules through the mucosal membrane, limited surface area, and short residence time limitations [59]. Hydrogels have increased residence time because of unique properties such as mucoadhesive and viscoelastic behavior. Therefore, hydrogels are used for delivering drugs by the nasal cavity for local diseases, intranasal vaccination, nasal congestion, allergic rhinitis, blood, and the brain [60]. The nasal mucosa is an attractive site for delivery as it is a painless and noninvasive method to deliver the drugs to the brain targeting. The nasal mucosa has received a lot of interest because it provides the potential for both mucosal and systemic immune responses, making it a desirable vaccination target area.

Thermoresponsive hydrogels are widely used for nasal delivery. A nasal congestion remedy was developed by encapsulating the phenylephrine hydrochloride in the hydrogel [61]. CS-based hydrogel doped with PEG–PLA nanoparticles was used to deliver with nucleic acid (miRNA-146) to treat allergic rhinitis [62]. The pharmacodynamic effects of miRNA release into the mucosa were evaluated, and the results were significantly better than other formulations. Plasmid-loaded CS nanoparticle-based hydrogel was used to treat hepatitis by nasal administration [63]. The efficacy of antidepressant therapy was enhanced by delivering berberine with hydroxypropyl-β-

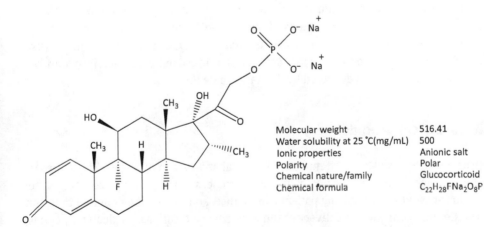

Molecular weight	516.41
Water solubility at 25 °C(mg/mL)	500
Ionic properties	Anionic salt
Polarity	Polar
Chemical nature/family	Glucocorticoid
Chemical formula	$C_{22}H_{28}FNa_2O_8P$

Figure 6.8: Structure and properties of dexamethasone 21-disodium phosphate (DXP) (Copyright 2019, reproduced after the permission of Elsevier [49]).

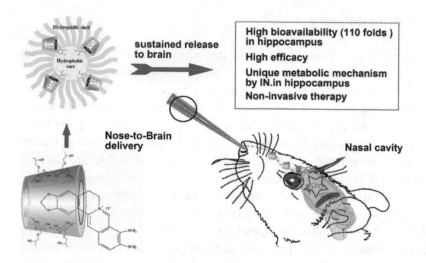

Figure 6.9: Schematic illustration for the nose-to-brain delivery of berberine/hydroxypropyl-β-cyclodextrin inclusion complex (Copyright 2019, reproduced after the permission of pubs.rsc.org [64]).

cyclodextrin through the nasal cavity [64]. Thermoresponsive hydrogel poloxamers (P407/P188, 16:2, w/w) formed an inclusion complex in hydrophilic berberine as shown in Figure 6.9.

The delivery of medications to the central nervous system is a significant obstacle in the treatment of disorders of the neurological system like meningitis, Parkinson's disease, Alzheimer's disease, and schizophrenia. Nasal drug delivery helps to deliver drugs by surpassing BBB. However, the constant flux of the clearance mechanism in the nose reduces the bioavailability of the drug. Nanostructured lipid carrier (NLC)-based hydrogels give an optimized solution for drug delivery for Alzheimer's disease [59]. Using Pluronic hydrogels as a nasal medication delivery technology, rivastigmine tartrate effectively targets the brain [65]. In another approach, thermosensitive hydrogel Pluronic was loaded with lorazepam having improved bioavailability, and sustained drug release was delivered by the nasal route [66].

6.3.3 Buccal drug delivery

Drug delivery through buccal mucosa has a local and systematic effect. This approach offers some advantages, such as minimising contact with digestive juices, preventing the first-pass effect, enhancing patient compliance, and having the option to time-control the treatment. For the delivery of some specific medications, particularly peptides and proteins, buccal drug delivery is therefore an interesting study area. For buccal drug delivery, a variety of dosage forms have been used, including films, patches, tablets, microparticles, and hydrogels. The most widely utilized dosage form for the buccal distribution of both hydrophilic and hydrophobic medications is mucoadhesive hydrogel films.

Thermoresponsive hydrogel for salbutamol delivery in the buccal area was prepared from poloxamer analogs (Kolliphor® P407/P188), xanthan gum (Satiaxane-® UCX930), and NaCl [67]. Using either chitosan glutamate (CHG), a soluble salt of CS, or a binary combination of CHG and glycerin, the anesthetic medication lidocaine hydrochloride (LDC) was administered in the buccal area [68]. Mucoadhesive polymers such as CS functionalized with catechol were developed for delivering LDC. Results confirmed that sustained release of the drug continued for about 3 h in rabbit buccal mucosa, without any inflammatory effects in the buccal tissue. Catechol was found to be an effective functionalization, leading to enhanced mucoadhesion of hydrogel [69]. Nanoparticle hydrogel composites were used to encapsulate drugs such as metoclopramide, ornidazole, insulin, and recombinant human epidermal growth factor, for buccal drug delivery [70].

Hydrogels have found their applications in the delivery of hydrophobic drugs as well in the buccal area. Carbopol polymers were used to prepare a mucoadhesive hydrogel containing ketoprofen, an anti-inflammatory and analgesic drug [71]. Carboxy-methyl-hexanoyl chitosan, β-glycerol phosphate, and glycerol hydrogel at 37 °C and pH 5.5 to 6.5 helped in the treatment of periodontitis by delivering naringin (anti-inflammatory drug) via buccal route [72].

6.3.4 Oral drug delivery

Oral drug delivery is an ancient, widely accepted, favored, and ease of administration drug delivery route. Moreover, oral drug administration is noninvasive and has high patient compliance. Common dosage forms are in tablets, capsules, syrups, suspensions, and powders. The major challenges are limited permeation, difficulty in disintegration, choking, and physiochemical instability [73]. Hydrogels are an interesting choice because of the optimized release of drugs, tunable properties, and biocompatibility. Therefore, hydrogels have a distinct supremacy in chemical and biopharmaceuticals.

Gajra et al. developed a hydrogel film by using PVA with the drug econazole nitrate to inhibit the growth of *Candida albicans* to cure oral candidiasis with a release time of 12 h [74]. The pH-responsive hydrogel consisting of polycaprolactone grafted on MAA copolymer was designed for the oral release of amifostine in the oral cavity [75]. This hydrogel provided radioprotection against acute radiation syndrome. Carbohydrate-based hydrogels were used to deliver drugs and biopharmaceuticals such as ornidazole, thymol, propranolol, flurbiprofen, ovalbumin, curcumin, progesterone, vancomycin, insulin, ciprofloxacin, 5-aminosalicylic acid, and ibuprofen by oral administration [76]. For instance, hyaluronic acid (HA) microparticle hydrogel was used to release vancomycin [77]. Sodium carboxymethyl cellulose was cross-linked with PVA in the ferric ion for flurbiprofen loading and drug release [78]. Carboxymethyl β-cyclodextrin-grafted carboxymethyl chitosan hydrogels were used to transport insulin through the oral route as shown in Figure 6.10 [79]. DNA hydrogels by microemulsification of cytosine-phosphate-guanine were coated by CS for an orally deliverable vaccine [80]. Treatment of various gastrointestinal tract diseases can be cured by targeted delivery of probiotics [81].

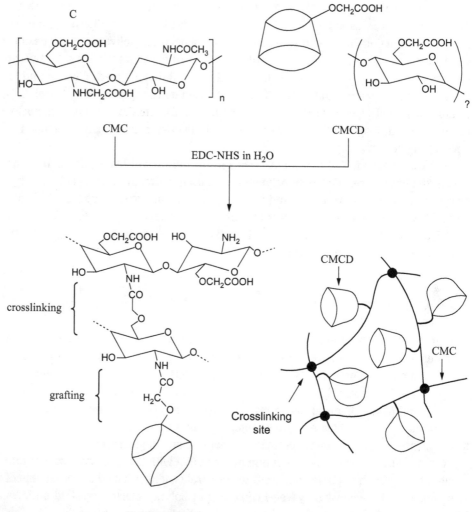

CMCD-g-CMC

Figure 6.10: Scheme for hydrogel synthesis and a schematic illustration of the structure of CMCD-g-CMC hydrogels (Copyright 2020, reproduced after the permission of Elsevier Ltd [79]).

6.3.5 Rectal drug delivery

The noninvasive medication administration technique of rectal drug delivery enables both local and systemic effects. The primary purposes of traditional rectal dose forms have been to give laxatives, administer antipyretics, and treat hemorrhoids for their local effects [82]. However, patients feel uncomfortable because of foreign body sensations. So, the development of drug delivery vehicles is an important area of research. Thermoresponsive hydrogels help in delivering drugs for various health problems such as cancer, pain killers, and antihypertensive drugs. The rectal route has the advantage of circumventing drug leakage as well as precise dosing [83].

A major challenge with cancer treatment drugs is to limit the toxic side effects. Rectal drug dispensing minimizes the toxicity of the drug. Thermoresponsive hydrogel made up of poloxamer P407 and 5-fluorouracil is a widely used therapy for colorectal cancer [84]. Similarly, oxiplatin suppository with Pluronic–poly(acrylic acid) (PAA) showed in situ gelling with convenient and effective rectal dosage form [85]. A combination of thermo-sensitive and pH-sensitive hydrogel made up of Pluronic and PAA was used to treat colon cancer by using epirubicin [86]. To reduce the side effects of irinotecan for the treatment of rectum cancer, irinotecan was encapsulated in solid lipid nanoparticles in a poloxamer solution. Additionally, docetaxel has been placed into a thermo-responsive hydrogel, whose rheological characteristics have undergone systemic evaluation [87].

Pain killers also known as analgesics are also delivered via the rectal route to avoid the side effects and high first-pass effects. As a result, these types of medications have also been administered rectally while enclosed in hydrogels. For instance, the acetaminophen-loaded thermo-responsive hydrogel was produced, administered via rectum in rats for at least 6 h without leakage [88]. Nimesulide was added in hydroxypropyl methylcellulose and sodium alginate with poloxamer to prepare mucoadhesive and thermosensitive hydrogel for the rectal therapy route [89]. The pharmacokinetics of antihypertensive drugs can be enhanced by rectal drug delivery. For example, propranolol was enclosed in a variety of hydrogels for rectal drug delivery. Furthermore, sodium alginate and sodium polycarbonate showed the maximum levels of mucus adhesion and the minimum levels of intrarectal migration, resulting in the highest bioavailability of propranolol (84.7% and 82.3%, respectively) [90].

6.3.6 Transdermal drug delivery

Drugs can be administered through the skin to have local or systemic effects using a transdermal drug delivery method as an effective alternative to oral and intravenous administration. In comparison to alternative administration methods, it has several benefits, including great patient compliance, noninvasive delivery, avoidance of gastrointestinal degradation, and the prevention of the first-pass effect. However, the use of transdermal medication delivery is constrained because the skin's corneum layer prevents effective drug penetration. Hydrogels used for transdermal drug routes are of two kinds: one is the common hydrogels and the second is the microneedle-based hydrogels.

Electrosensitive hydrogel consisting of polyacrylamide-grafted-pectin copolymer was used as a transdermal drug delivery system for rivastigmine delivery [91]. Gelatin and polyacrylamide-based stretchable and tough hydrogel for the release of diltiazem, nicotine, diclofenac, and lidocaine was evaluated as a transdermal therapy [92]. Zinc oxide nanorods with CS were used to treat cardiovascular system diseases by using acetylsalicylic acid as a transdermal therapy [93].

Microneedles are micro level medical devices that contain hundreds of individual micro-needles, with each needle's length varying from 25 to 1,000 µm. Microneedles have drugs in polymer placed in reservoirs attached to the hydrogel-based microneedle For example, rifampicin, isoniazid, pyrazinamide, and ethambutol were all delivered transdermally using a new hydrogel-forming microneedle array. These medications were added to several kinds of drug reservoirs, and for improved drug delivery, they were also combined with microneedle arrays that form hydrogels [94]. MAA ester-based microneedles for transdermal infusion of drugs were designed for drug delivery [95].

6.3.7 Subcutaneous and intramuscular (injection) drug delivery

Subcutaneous (under the skin) and intramuscular (in a muscle) medication delivery are the two basic types of injectable administration, sometimes known as parenteral administration [96]. Drugs are injected into a person's subcutaneous layer, which is the layer between the skin and the muscle, and it is frequently done for diabetes and cancer patients. Healthcare professionals and patients alike choose to use this method of drug delivery because it has been shown to be secure, well-tolerated, and efficient, particularly for the administration of many protein-based medications [97]. The dose for subcutaneous medication delivery is only allowed to be 2 mL, though. When more medicine is needed in a given amount, intramuscular drug administration is used over subcutaneous drug delivery. It is also referred to as an intramuscular injection because the medicine enters the deepest layer of the muscle through the dermis and subcutaneous tissues, where the robust blood supply enables quick and complete absorption [98]. Both subcutaneous and intramuscular medication deliveries have been carried out using hydrogel as the delivery system.

For instance, donepezil was added to a hydrogel based on polypseudorotaxane structure and polydopamine bond-based cross-linked HA hydrogel for subcutaneous injection [99]. After subcutaneous injection, donepezil can be released from the hydrogel over an extended period. The same study team also suggested another method of injecting an HA hydrogel subcutaneously. Alzheimer's disease treatment was developed by injecting HA hydrogel (microstructured lipid carriers/hyaluronic acid hydrogel) with acetylcholinesterase daily [100].

6.4 Conclusion

In this chapter, the classification of hydrogel according to the source/origin, polymeric composition, structure, and drug delivery has been summarized. The fabrication strategies of the hydrogel by the advanced techniques and physical and chemical methods have been found effective for drug delivery. In addition, the applications of hydrogels to cure diseases by carrying drugs, biological molecules, and cells to the target site by various routes have been discussed. Polymeric systems such as monomers and copolymers for efficient drug delivery vehicles are described. Various polymers and biomaterials having optimal mechanical strength, enhanced biocompatibility, tunable biodegradability, minimal toxicity, and favorable swelling behavior for drug delivery applications in various administration routes have been discussed.

References

[1]   Lin, X., et al. (2022). Progress in the mechanical enhancement of hydrogels: Fabrication strategies and underlying mechanisms. J Polym Sci, 60(17), 2525–2542.

[2]   Wichterle, O., & Lim, D. (1960). Hydrophilic gels for biological use. Nature, 185(4706), 117–118.

[3]   Hoffman, A. S. (2012). Hydrogels for biomedical applications. Adv Drug Deliv Rev, 64, 18–23.

[4]   Lim, F. & Sun, A. M. (1980). Microencapsulated islets as the bioartificial endocrine pancreas. Science, 210(4472), 908–910.

[5]   Yannas, I., et al. (1989). Synthesis and characterization of a model extracellular matrix that induces partial regeneration of adult mammalian skin. Proc Natl Acad Sci, 86(3), 933–937.

[6]   Ekenseair, A. K., Kasper, F. K., & Mikos, A. G. (2013). Perspectives on the interface of drug delivery and tissue engineering. Adv Drug Deliv Rev, 65(1), 89–92.

[7]   Sennakesavan, G., et al. (2020). Acrylic acid/acrylamide based hydrogels and its properties-A review. Polym Degrad Stab, 180, 109308.

[8]   Tamburini, G., et al. (2022). Optimized semi-interpenetrated p (HEMA)/PVP hydrogels for artistic surface cleaning. Materials, 15(19), 6739.

[9]   Li, J., et al. (2020). The Potential of electrospinning/electrospraying technology in the rational design of hydrogel structures. Macromol Mater Eng, 305(8), 2000285.

[10]   Luo, Y., Kirker, K. R., & Prestwich, G. D. (2000). Cross-linked hyaluronic acid hydrogel films: New biomaterials for drug delivery. J Control Release, 69(1), 169–184.

[11]   Patwekar, S. L. (2016). Nanocomposite: A new approach to drug delivery system. Asian J Pharm (AJP), 10(04).

[12] Oh, S. H., et al. (2003). Fabrication and characterization of hydrophilic poly (lactic-co-glycolic acid)/ poly (vinyl alcohol) blend cell scaffolds by the melt-molding particulate-leaching method. Biomaterials, 24(22), 4011–4021.

[13] Okamoto, M. & John, B. (2013). Synthetic biopolymer nanocomposites for tissue engineering scaffolds. Prog Polym Sci, 38(10–11), 1487–1503.

[14] Reddi, A. H. (1998). Role of morphogenetic proteins in skeletal tissue engineering and regeneration. Nat Biotechnol, 16(3), 247–252.

[15] Patel, V. R. & Amiji, M. M. (1996). Preparation and characterization of freeze-dried chitosan-poly (ethylene oxide) hydrogels for site-specific antibiotic delivery in the stomach. Pharm Res, 13, 588–593.

[16] Niculescu, A.-G., et al. (2021). Fabrication and applications of microfluidic devices: A review. Int J Mol Sci, 22(4), 2011.

[17] Zhao, Z., et al. (2021). Injectable microfluidic hydrogel microspheres for cell and drug delivery. Adv Funct Mater, 31(31), 2103339.

[18] Gupta, P. & Purwar, R. (2020). Electrospun pH responsive poly (acrylic acid-co-acrylamide) hydrogel nanofibrous mats for drug delivery. J Polym Res, 27(10), 296.

[19] Kai, D., Liow, S. S., & Loh, X. J. (2014). Biodegradable polymers for electrospinning: Towards biomedical applications. Mater Sci Eng C, 45, 659–670.

[20] Ma, X., et al. (2014). Nanofibrous electroactive scaffolds from a chitosan-grafted-aniline tetramer by electrospinning for tissue engineering. RSC Adv, 4(26), 13652–13661.

[21] Bao, M., et al. (2014). Electrospun biomimetic fibrous scaffold from shape memory polymer of PDLLA-co-TMC for bone tissue engineering. ACS Appl Mater Interfaces, 6(4), 2611–2621.

[22] Ahadi, F., Khorshidi, S., & Karkhaneh, A. (2019). A hydrogel/fiber scaffold based on silk fibroin/ oxidized pectin with sustainable release of vancomycin hydrochloride. Eur Polym J, 118, 265–274.

[23] Yohe, S. T., Colson, Y. L., & Grinstaff, M. W. (2012). Superhydrophobic materials for tunable drug release: Using displacement of air to control delivery rates. J Am Chem Soc, 134(4), 2016–2019.

[24] Ghosh, T., Das, T., & Purwar, R. (2021). Review of electrospun hydrogel nanofiber system: Synthesis, properties and applications. Polym Eng Sci, 61(7), 1887–1911.

[25] Xu, J., et al. (2011). Controlled dual release of hydrophobic and hydrophilic drugs from electrospun poly (l-lactic acid) fiber mats loaded with chitosan microspheres. Mater Lett, 65(17–18), 2800–2803.

[26] Hou, Z., et al. (2013). Electrospun upconversion composite fibers as dual drugs delivery system with individual release properties. Langmuir, 29(30), 9473–9482.

[27] Revzin, A., et al. (2001). Fabrication of poly (ethylene glycol) hydrogel microstructures using photolithography. Langmuir, 17(18), 5440–5447.

[28] Bose, S., Vahabzadeh, S., & Bandyopadhyay, A. (2013). Bone tissue engineering using 3D printing. Mater Today, 16(12), 496–504.

[29] Fitzgerald, S. (2015). FDA approves first 3D-printed epilepsy drug experts assess the benefits and caveats. Neurol Today, 15(18), 26–27.

[30] Khoeini, R., et al. (2021). Natural and synthetic bioinks for 3D bioprinting. Adv NanoBiomed Res, 1 (8), 2000097.

[31] Nasrollahi, F., et al. (2021). Graphene quantum dots for fluorescent labeling of gelatin-based shear-thinning hydrogels. Adv NanoBiomed Res, 1(7), 2000113.

[32] Nazir, F. & Iqbal, M. (2022). Piezoelectric MoS2 nanoflowers (NF's) for targeted cancer therapy by gelatin-based shear thinning hydrogels. React Funct Polym, 181, 105435.

[33] Xu, J., Liu, Y., & Hsu, S.-H. (2019). Hydrogels based on Schiff base linkages for biomedical applications. Molecules, 24(16), 3005.

[34] Samaddar, P., Kumar, S., & Kim, K. -. H. (2019). Polymer hydrogels and their applications toward sorptive removal of potential aqueous pollutants. Polym Rev, 59(3), 418–464.

[35] Coviello, T., et al. (1999). Novel hydrogel system from scleroglucan: Synthesis and characterization. J Control Release, 60(2–3), 367–378.

[36] Bencherif, S. A., et al. (2009). Nanostructured hybrid hydrogels prepared by a combination of atom transfer radical polymerization and free radical polymerization. Biomaterials, 30(29), 5270–5278.

[37] Baloch, A., et al. (2022). Fabrication of swellable PEGylated hydrogel by free radical polymerization for controlled delivery of non-steroidal anti-inflammatory drug; characterization and statistical analysis. Part Sci Technol, 1–14.

[38] Zou, D., et al. (2022). Boric acid-loosened polyvinyl alcohol/glutaraldehyde membrane with high flux and selectivity for monovalent/divalent salt separation. J Membr Sci, 662, 120954.

[39] Reghioua, A., et al. (2021). Magnetic chitosan-glutaraldehyde/zinc oxide/Fe3O4 nanocomposite: Optimization and adsorptive mechanism of remazol brilliant blue R dye removal. J Polym Environ, 29(12), 3932–3947.

[40] Nazir, F., et al. (2021). 6-deoxy-aminocellulose derivatives embedded soft gelatin methacryloyl (GelMA) hydrogels for improved wound healing applications: In vitro and in vivo studies. Int J Biol Macromol, 185, 419–433.

[41] Piao, Y., et al. (2021). Biomedical applications of gelatin methacryloyl hydrogels. Eng Regen, 2, 47–56.

[42] Han, Y., et al. (2021). Biomimetic injectable hydrogel microspheres with enhanced lubrication and controllable drug release for the treatment of osteoarthritis. Bioactive Mater, 6(10), 3596–3607.

[43] Nazir, F., et al. (2021). Fabrication of robust poly l-lactic acid/cyclic olefinic copolymer (PLLA/COC) blends: Study of physical properties, structure, and cytocompatibility for bone tissue engineering. J Mater Res Technol, 13, 1732–1751.

[44] De Oliveira Cardoso, V. M., et al. (2017). Development and characterization of cross-linked gellan gum and retrograded starch blend hydrogels for drug delivery applications. J Mech Behav Biomed Mater, 65, 317–333.

[45] Ghauri, Z. H., et al. (2021). Development and evaluation of pH-sensitive biodegradable ternary blended hydrogel films (chitosan/guar gum/PVP) for drug delivery application. Sci Rep, 11(1), 21255.

[46] Schultz, C., et al. (2011). Drug delivery to the posterior segment of the eye through hydrogel contact lenses. Clin Exp Optom, 94(2), 212–218.

[47] Cooper, R. C. & Yang, H. (2019). Hydrogel-based ocular drug delivery systems: Emerging fabrication strategies, applications, and bench-to-bedside manufacturing considerations. J Control Release, 306, 29–39.

[48] Ali, F., et al. (2022). Emerging fabrication strategies of hydrogels and its applications. Gels, 8(4), 205.

[49] Faccia, P. A., Pardini, F. M., & Amalvy, J. I. (2019). Uptake and release of Dexamethasone using pH-responsive poly (2-hydroxyethyl methacrylate-co-2-(diisopropylamino) ethyl methacrylate) hydrogels for potential use in ocular drug delivery. J Drug Deliv Sci Technol, 51, 45–54.

[50] Almeida, H., et al. (2016). Development of mucoadhesive and thermosensitive eye drops to improve the ophthalmic bioavailability of ibuprofen. J Drug Deliv Sci Technol, 35, 69–80.

[51] Shastri, D. H., Prajapati, S. T., & Patel, L. D. (2010). Thermoreversible mucoadhesive ophthalmic in situ hydrogel: Design and optimization using a combination of polymers. Acta Pharm, 60(3), 349–360.

[52] Wang, A., et al. (2022). A methacrylated hyaluronic acid network reinforced Pluronic F-127 gel for treatment of bacterial keratitis. Biomed Mater.

[53] Cheng, Y.-H., et al. (2019). Thermosensitive chitosan-gelatin-based hydrogel containing curcumin-loaded nanoparticles and latanoprost as a dual-drug delivery system for glaucoma treatment. Exp Eye Res, 179, 179–187.

[54]   Grimaudo, M. A., et al. (2020). Micelle-nanogel platform for ferulic acid ocular delivery. Int J pharm, 576, 118986.

[55]   Xu, X., et al. (2020). Functional chitosan oligosaccharide nanomicelles for topical ocular drug delivery of dexamethasone. Carbohydr Polym, 227, 115356.

[56]   Yadav, S., et al. (2022). Preparation and in-vitro characterization of brimonidine encapsulated polymer coated microemulsion for ocular drug delivery for the management of glaucoma. J Harbin Inst Technol, 54(5), 2022.

[57]   Li, B., et al. (2020). Drug-loaded chitosan film prepared via facile solution casting and air-drying of plain water-based chitosan solution for ocular drug delivery. Bioactive Mater, 5(3), 577–583.

[58]   Yu, Y., et al. (2021). Recent advances in thermo-sensitive hydrogels for drug delivery. J Mater Chem B, 9(13), 2979–2992.

[59]   Cunha, S., et al. (2021). Improving drug delivery for Alzheimer's disease through nose-to-brain delivery using nanoemulsions, nanostructured lipid carriers (NLC) and in situ hydrogels. Int J Nanomed, 16, 4373.

[60]   Salatin, S., et al. (2016). Hydrogel nanoparticles and nanocomposites for nasal drug/vaccine delivery. Arch Pharm Res, 39(9), 1181–1192.

[61]   Xu, X., et al. (2014). Preparation and in vitro characterization of thermosensitive and mucoadhesive hydrogels for nasal delivery of phenylephrine hydrochloride. Eur J Pharm Biopharm, 88(3), 998–1004.

[62]   Su, Y., et al. (2020). Chitosan hydrogel doped with PEG-PLA nanoparticles for the local delivery of miRNA-146a to treat allergic rhinitis. Pharmaceutics, 12(10), 907.

[63]   Khatri, K., et al. (2008). Plasmid DNA loaded chitosan nanoparticles for nasal mucosal immunization against hepatitis B. Int J pharm, 354(1–2), 235–241.

[64]   Wang, Q.-S., et al. (2020). Intranasal delivery of berberine via in situ thermoresponsive hydrogels with non-invasive therapy exhibits better antidepressant-like effects. Biomater Sci, 8(10), 2853–2865.

[65]   Abouhussein, D. M., et al. (2018). Brain targeted rivastigmine mucoadhesive thermosensitive In situ gel: Optimization, in vitro evaluation, radiolabeling, in vivo pharmacokinetics and biodistribution. J Drug Deliv Sci Technol, 43, 129–140.

[66]   Jose, S., et al. (2013). Thermo-sensitive gels containing lorazepam microspheres for intranasal brain targeting. Int J pharm, 441(1–2), 516–526.

[67]   Zeng, N., et al. (2014). Influence of additives on a thermosensitive hydrogel for buccal delivery of salbutamol: Relation between micellization, gelation, mechanic and release properties. Int J pharm, 467(1–2), 70–83.

[68]   Pignatello, R., Basile, L., & Puglisi, G. (2009). Chitosan glutamate hydrogels with local anesthetic activity for buccal application. Drug Deliv, 16(3), 176–181.

[69]   Xu, J., et al. (2015). Genipin-crosslinked catechol-chitosan mucoadhesive hydrogels for buccal drug delivery. Biomaterials, 37, 395–404.

[70]   Morantes, S. J., et al. (2017). Composites of hydrogels and nanoparticles: A potential solution to current challenges in buccal drug delivery. In Biopolymer-Based Composites (pp. 107–138). Elsevier.

[71]   Özyazıcı, M., et al. (2015). Bioadhesive gel and hydrogel systems for buccal delivery of ketoprofen: Preparation and in vitro evaluation studies. biosensors, 6(7).

[72]   Chang, P. C., et al. (2017). Inhibition of periodontitis induction using a stimuli-responsive hydrogel carrying naringin. J Periodontol, 88(2), 190–196.

[73]   Basha, S. K., et al. (2021). Solid lipid nanoparticles for oral drug delivery. Mater Today Proc, 36, 313–324.

[74]   Gajra, B., et al. (2014). Mucoadhesive hydrogel films of econazole nitrate: Formulation and optimization using factorial design. J Drug Deliv, 2014.

[75]   Lin, X., et al. (2020). Design and evaluation of pH-responsive hydrogel for oral delivery of amifostine and study on its radioprotective effects. Colloids Surf B Biointerfaces, 195, 111200.

[76]   Ahadian, S., et al. (2020). Micro and nanoscale technologies in oral drug delivery. Adv Drug Deliv Rev, 157, 37–62.

[77]   Sahiner, N., Suner, S. S., & Ayyala, R. S. (2019). Mesoporous, degradable hyaluronic acid microparticles for sustainable drug delivery application. Colloids Surf B Biointerfaces, 177, 284–293.

[78]   Bulut, E. (2020). Chitosan coated-and uncoated-microspheres of sodium carboxymethyl cellulose/polyvinyl alcohol crosslinked with ferric ion: Flurbiprofen loading and in vitro drug release study. J Macromol Sci, 57(1), 72–82.

[79]   Yang, Y., et al. (2020). Carboxymethyl β-cyclodextrin grafted carboxymethyl chitosan hydrogel-based microparticles for oral insulin delivery. Carbohydr Polym, 246, 116617.

[80]   Nomura, D., et al. (2018). Development of orally-deliverable DNA hydrogel by microemulsification and chitosan coating. Int J pharm, 547(1–2), 556–562.

[81]   Enck, K., et al. (2020). Development of a novel oral delivery vehicle for probiotics. Curr Pharm Des, 26(26), 3134–3140.

[82]   Surti, N., Mahajan, A., & Misra, A. (2021). Polymers in rectal drug delivery. In Applications of Polymers in Drug Delivery (pp. 263–280). Elsevier.

[83]   Bialik, M., et al. (2021). Achievements in thermosensitive gelling systems for rectal administration. Int J Mol Sci, 22(11), 5500.

[84]   Kassab, H. J., & Khalil, Y. I. (2014). 5–Fluorouracil mucoadhesive liquid suppository formulation and evaluation. World J Pharm Res, 3(9), 119–135.

[85]   Lin, H.-R., et al. (2012). A novel in-situ-gelling liquid suppository for site-targeting delivery of anti-colorectal cancer drugs. J Biomater Sci Polym Ed, 23(6), 807–822.

[86]   Lo, Y.-L., Lin, Y., & Lin, H. –. R. (2013). Evaluation of epirubicin in thermogelling and bioadhesive liquid and solid suppository formulations for rectal administration. Int J Mol Sci, 15(1), 342–360.

[87]   Yeo, W. H., et al. (2013). Docetaxel-loaded thermosensitive liquid suppository: Optimization of rheological properties. Arch Pharm Res, 36, 1480–1486.

[88]   Choi, H.-G., et al. (1998). Development of in situ-gelling and mucoadhesive acetaminophen liquid suppository. Int J pharm, 165(1), 33–44.

[89]   Yuan, Y., et al. (2012). Thermosensitive and mucoadhesive in situ gel based on poloxamer as new carrier for rectal administration of nimesulide. Int J pharm, 430(1–2), 114–119.

[90]   Ryu, J. –. M., et al. (1999). Increased bioavailability of propranolol in rats by retaining thermally gelling liquid suppositories in the rectum. J Control Release, 59(2), 163–172.

[91]   Patil, S. B., et al. (2020). Functionally tailored electro-sensitive poly (acrylamide)-g-pectin copolymer hydrogel for transdermal drug delivery application: Synthesis, characterization, in-vitro and ex-vivo evaluation. Drug Deliv Lett, 10(3), 185–196.

[92]   Qiao, Z., et al. (2021). Highly stretchable gelatin-polyacrylamide hydrogel for potential transdermal drug release. Nano Select, 2(1), 107–115.

[93]   Radwan-Pragłowska, J., et al. (2020). ZnO nanorods functionalized with chitosan hydrogels crosslinked with azelaic acid for transdermal drug delivery. Colloids Surf B Biointerfaces, 194, 111170.

[94]   Anjani, Q. K., et al. (2021). Versatility of hydrogel-forming microneedles in in vitro transdermal delivery of tuberculosis drugs. Eur J Pharm Biopharm, 158, 294–312.

[95]   Yeung, C., et al. (2019). A 3D-printed microfluidic-enabled hollow microneedle architecture for transdermal drug delivery. Biomicrofluidics, 13(6), 064125.

[96]   Schou, J. (1971). Subcutaneous and intramuscular injection of drugs. Concepts Biochem Pharmacol, 1, 47–66.

[97]   Li, C., et al. (2021). Advances in subcutaneous delivery systems of biomacromolecular agents for diabetes treatment. Int J Nanomed, 16, 1261.

[98] Bellotti, E., et al. (2021). Injectable thermoresponsive hydrogels as drug delivery system for the treatment of central nervous system disorders: A review. J Control Release, 329, 16–35.

[99] Hwang, C., et al. (2021). Polypseudorotaxane and polydopamine linkage-based hyaluronic acid hydrogel network with a single syringe injection for sustained drug delivery. Carbohydr Polym, 266, 118104.

[100] Kang, N.-W., et al. (2021). Subcutaneously injectable hyaluronic acid hydrogel for sustained release of donepezil with reduced initial burst release: Effect of hybridization of microstructured lipid carriers and albumin. Pharmaceutics, 13(6), 864.

Zubair Ahmad*, Tahseen Arshad, Hassan Zeb, Shahid Ali Khan*

Chapter 7
Hydrogels for personal care products

Abstract: The hydrophilic polymer networks that make up hydrogels are capable of absorbing large quantity of water, which increases their volume and causes them to exhibit a wide range of diverse material characteristics. They have been used in a variety of biological sectors since their first practical application in the 1960s. After more than 50 years of industrial usage, a variety of hydrogels are now available on the market for various uses and with a wide range of properties. Even though it would be nearly impossible to identify all the commercial products based on hydrogels for biological applications, a thorough study of the materials that have made it to market has been done in this chapter. Moreover, the handling of the menstrual cycle via unhygienic cloth-based products and their subsequent health problems are also discussed here. However, the primary aim of this chapter is to target the application of hydrogels to personal care products, including baby diapers and feminine incontinence products like sanitary pads, and to describe their advantages. Aside from these, this chapter covers a brief history, the structure, and classification of hydrogels.

7.1 Introduction

The word "hydrogel" first appeared in 1894 when it was employed to describe colloidal gels made from specific inorganic salts [1]. The name "hydrogel" today refers to a class of materials that is quite different from what it originally meant: hydrogels are three-dimensional cross-linked networks of polymer chains that have the ability to absorb and

Acknowledgments: The Authors highly acknowledge the Department of Chemistry, School of Natural Sciences, National University of Science and Technology (NUST), Islamabad 44000, Pakistan, and the Department of Chemistry, University of Swabi, Swabi for providing necessary facility to this work.

*Corresponding author: Zubair Ahmad**, Department of Chemistry, University of Swabi, Anbar, Khyber Pakhtunkhwa 23561, Pakistan, e-mail: za3724364@gmail.com
*Corresponding author: Shahid Ali Khan**, Department of Chemistry, School of Natural Sciences, National University of Science and Technology (NUST), Islamabad 44000, Pakistan;
Department of Urology, Key Laboratory of Biological Targeting Diagnosis, Therapy and Rehabilitation of Guangdong Higher Education Institutes, The Fifth Affiliated Hospital of Guangzhou Medical University, Guangzhou Medical University, Guangzhou 510700, China, e-mail: shahid.ali@sns.nust.edu.pk
Tahseen Arshad, Department of Chemistry, School of Natural Sciences, National University of Science and Technology (NUST), Islamabad 44000, Pakistan
Hassan Zeb, Department of Statistics, Islamia College University, Peshawar, Khyber Pakhtunkhwa 25120, Pakistan

https://doi.org/10.1515/9783111334080-007

retain significant quantities of water in the gaps among chains. A hydrogel containing poly(vinyl alcohol) cross-linked with formaldehyde and sold under the brand name Ivalon, used for biomedical implants, had its commercial debut in 1949 [2].

The first and effective utilization hydrogels in contact lenses has led to a wide range of hydrogel uses today. In tissue of engineering, hydrogels are now employed as scaffolds and may include cells that can mend damaged tissues. Hydrogels that are responsive to the environment can detect changes in pH, temperatures, or metabolite concentration and release their content as a response. Hydrogels with particular molecular responsiveness, such as glucose or antigens, can be employed as sensing devices and as restricted delivery mechanisms for agrochemicals and bioactive compounds [3, 4].

The majority of commercial superabsorbent polymer (SAP) use is for the urine and blood-absorbent SAP hydrogels found in hygiene goods including feminine incontinence products and newborn diapers. This chapter tries to address nonhygienic SAP uses as there are multiple review articles on hygienic usage of SAP hydrogels [2].

According to a historical categorization, hydrogels may be categorized into many eras.

The primary method used to create the first generation of hydrogels was a chain addition process, which typically involves vinyl monomers and a starting free radical molecule. Free radical polymerization continues until two radical species recombine or line divided occurs. Polyacrylamide, one of the principal hydrogel-forming polymers has been used in industrial settings as an agronomic gel. Poly(hydroxyalkyl methacrylate), another significant polymer, is still a crucial component in the creation of contact lenses even though it was discovered more than 50 years ago [5]. The second main group of the first-generation hydrogels consists of hydrophilic polymers that are covalently bonded. Polyvinyl alcohol (PVA) and polyethylene glycol (PEG) are two of them that continue to be crucial in tissue engineering.

The stimulus-sensitive hydrogels, the second generation hydrogels, were released in the market in the 1970s. Utilizing these hydrogels made it possible to activate a particular event in reaction to an alteration in the environment, like the release of a medication or the erosion of a polymer (i.e., temperature or pH). PEG-polyester block copolymers, sometimes referred to as poloxamers or Pluronics (by BASF), must be recognized as being a part of this generation of hydrogels. These polymers have a phase transition property that allows them to go from the sol to the gel state at relatively low temperature, and from gel to the sol state at higher temperatures. This property has been extensively employed in controlled and sustained release [6].

Nowadays, hydrogels are using to make microcapsules and microparticles for use in cosmetic and medicinal procedures. Hydrogels are mostly administered topically to the skin, hair, and mouth in cosmetic purposes. Long residence periods on the application site and a reduced need for product administration frequency are only two of the significant benefits of using bioadhesive hydrogels for skin care applications. As of now, various cosmetic formulations that comprise active cosmetic components have been created from hydrogels. The chosen hydrogels have appropriate bioadhesive compositions for skin care products [7–9].

As they are able to retain moisture away from skin, encouraging skin health, avoiding diaper rash, and giving comfort, superabsorbent hydrogels – particularly those made of acrylate-based materials – are widely used in personal care products including feminine incontinence and baby diapers to absorb fluids [10, 11].

7.2 Classification of hydrogels

Hydrogels are categorize on the basis of source, polymeric composition, physical structure, and cross-linking [4] as shown in the inset of Figure 7.1.

7.2.1 Based on source

On the basis of source or origin, hydrogels are categorized into natural, synthetic, and semisynthetic/hybrid hydrogels.

7.2.1.1 Natural hydrogels

Gelatin and collagen hydrogels composed of natural polymers are commonly known as natural hydrogels. Natural hydrogels are widely distributed and have a variety of properties, including a large absorption rate, biodegradability, tunable porosity, and biocompatibility [12].

7.2.1.2 Synthetic hydrogels

Contrarily, synthetic polymers like polyamides and polyethylene glycol are used to synthesize synthetic hydrogels. Hydrophilic homopolymers or copolymers that have been ionically or covalently cross-linked form swelling, three-dimensional networks that make up synthetic polymeric hydrogels [13].

7.2.1.3 Hybrid hydrogels

Natural and synthetic polymer hydrogels are combined to prepare hybrid hydrogels. Collagen, chitosan, and dextran are examples of natural biopolymers that have been mixed with synthetic polymers like poly(*N*-isopropyl acrylamide) and PVA [14].

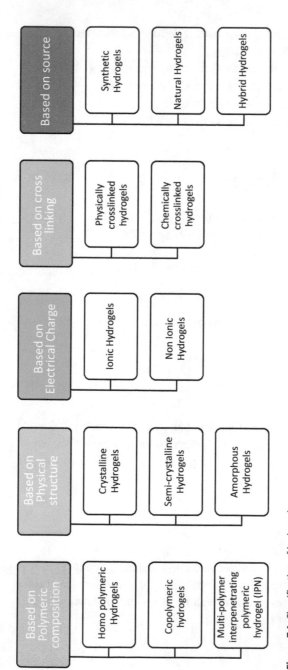

Figure 7.1: Classification of hydrogels.

7.2.2 Classification based on polymeric composition

Hydrogels are classified into three classes based on polymeric composition which is described in the below sections.

7.2.2.1 Homopolymeric hydrogels

Homopolymeric hydrogels are a type of polymer network, and composed of one monomer. Homopolymeric hydrogels are a fundamental structural component of all polymer networks hydrogels. Based on the kind of monomer and the polymerization process, homopolymers may have a cross-linked skeletal structure. An example of such hydrogels is poly(2-hydroxyethyl methacrylate) [15].

7.2.2.2 Copolymeric hydrogels

Typically, two or more different monomeric species that have at least one hydrophilic monomeric component are present in copolymeric hydrogels. These monomeric species are organized along the polymeric networking chains in a randomized, block, and/or consecutive structural arrangement. An example of such hydrogel is poly(vinyl pyrrolidone/acrylic acid) [16].

7.2.2.3 Multipolymer interpenetrating polymeric hydrogel

It is a particular type of hydrogel where the network of cross-linked polymers contains two or more polymeric units. Interpenetrating networks are viewed as cross-linked polymer "alloys," and unless the chemical bonds are broken, these networks cannot be divided. An example of such hydrogel is polyvinyl alcohol [17].

7.2.3 Classification based on physical structure

Hydrogels are further classified into three groups based on the physical structure.

7.2.3.1 Crystalline hydrogels

These hydrogels possess a crystalline structure, outstanding stability, and optimum resilience under various circumstances, such as acidic or basic conditions, or in the existence of charged biopolymers. It also has great rheological characteristics [18].

7.2.3.2 Semicrystalline hydrogels

Semicrystalline hydrogels are water-swelling hydrogels with crystalline domains that were first synthesized in 1994. According to the recent research, physically cross-linked semicrystalline hydrogels are among the mechanically robust and extremely stretchy hydrogels with functionalities such as melt processing, self-healing, and shape memory. They can rapidly and reversibly switch from a solid-like state to a liquid-like state when exposed to the melting temperature, thereby providing numerous potential applications [19].

7.2.3.3 Amorphous hydrogels

Glycerin and water-based compounds are called amorphous hydrogels, and are typically produced for hydrating wounds. These dressings support granulation and epithelialization, encourage wound healing in a wet environment, and make autolytic debridement easier [20].

7.2.4 Classification based on electrical charge

On the basis of electrical charge, hydrogels may be classified into four categories.

7.2.4.1 Ionic hydrogels

Ionic hydrogels are a type of hydrogel that contains charged ionic groups, such as carboxylate, sulfonate, or quaternary ammonium within their structure. These charged groups can interact with oppositely charged ions in the surrounding environment, resulting in swelling or shrinking of the hydrogel. Ionic hydrogels are commonly used in drug delivery, tissue engineering, and sensors applications [21].

7.2.4.2 Nonionic hydrogels

In non-ionic hydrogels the network can be homopolymeric or copolymeric. Such hydrogels are made without the use of charged groups. These hydrogels can be produced by using various polymerization procedures or manufacture along the polymer chain [22].

7.2.5 Classification based on cross-linking

In these hydrogels, the cross-linking either be physical or chemical.

7.2.5.1 Physically cross-linked hydrogels

Physical cross-linked hydrogels are a type of hydrogel that is held together by non-covalent interactions such as hydrogen bonding or van der Waals forces, rather than covalent bonds. These hydrogels are formed by the physical processes like heating or cooling, or by mixing two different polymers that have a complementary interaction to form a hydrogel network. Physical cross-linked hydrogels have unique properties such as high swelling capacity, mechanical stability, and biocompatibility, making them useful for a wide range of applications including drug delivery, tissue engineering, and wound healing. Owing to the weak interactions the physical cross-linked hydrogels are regarded as reversible hydrogels [23].

7.2.5.2 Chemically cross-linked hydrogels

Chemical cross-linking is the connecting of two or more molecules by a covalent bond either intramolecularly or intermolecularly. Cross-linking reagents or "cross-linkers" are the terms used to describe the reagents employed for the purpose. Covalent bonding between polymer chains is used in the chemical cross-linking process to create permanent hydrogel. The chemical cross-links in hydrogels can be achieved by using polymer-polymer conjugation, small cross-linker molecules, enzyme-catalyzed reactions, or photosensitive agents [24].

7.3 Application of hydrogels

Hydrogels can be created by physically or chemically cross-linking hydrophilic polymers, both natural and manufactured. Their similarity to live tissue creates a wide range of possibilities for biological applications. Currently, hydrogels are utilized to prepare wound healing, tissue engineering scaffolds, contact lenses, personal care products, and drug delivery systems [4, 7, 25, 26] as shown in Figure 7.2.

7.3.1 Hydrogels in personal hygiene products

Disposable nappies, sanitary napkins, and adult protective undergarments for incontinence are examples of hygiene products based on hydrogels [27]. SAPs, which are widely utilized to absorb fluids, are consumed most by disposable diapers among other items. Three-dimensional cross-linked polymers (linear or branched) are known as SAPs because of their exceptional hydrophilic capabilities, which enable them to absorb and retain large volumes of liquid. They can absorb up to 1,000 times their

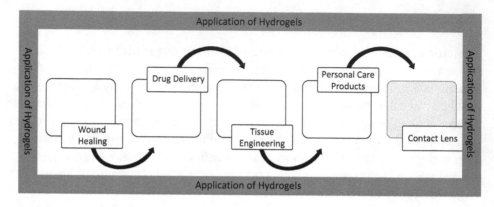

Figure 7.2: Application of hydrogels.

weight in water (or biological fluids), which gives them a great swelling capacity. The primary benefits of using these polymers are their ability to keep moisture away from the skin, which promotes skin health and prevents diaper rash and microbial colonization [28, 29].

Since 1970s, there has been a great deal of commercial attention in nappy technology, and during the past 25 years, around 1,000 diaper-related patents have been published. After Germany and France began employing them in infant diapers in 1980, Japan began the first commercial manufacturer of hydrogels in 1978 for use in feminine pads. In 1982, it was suggested in Japan that SAP may be used in the diaper sector to reduce diaper size (by roughly 50%), improve retention, and lower leakage values to around 2%. The most popular cross-linked synthetic polymers utilized today as superabsorbent materials include acrylic acid and also its copolymers with acrylamide [28, 30].

The generation of solid waste is one of the main downsides of diaper usage. In fact, according to reports, a child under the age of 30 months who uses disposable diapers roughly produces 1,092 m^3 of trash per anum. The third largest consumer item in landfills is disposable diapers, which account for 4% of all solid waste. About 50% of household trash in a home with a child in nappies comes from disposables. Thus, reuse of products like baby nappies, napkins, medical bedsheets, feminine towels, and other related items has been actively pursued as depicted in Figure 7.3. Such complicated goods are recycled by isolating the elements for different applications. Researchers began to consider other options, such as employing entirely biodegradable components. Because of the capillary effects, new forms of hydrogels include divinyl sulfone and sodium carboxymethyl cellulose, and hydroxyethyl cellulose may swell like SAPs and have significant water retention [28, 31].

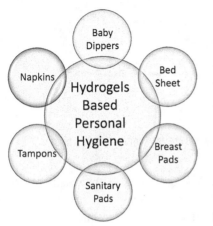

Figure 7.3: Hydrogel-based personal care products.

7.3.1.1 Hydrogels in baby diapers

When compared to cloth diapers, superabsorbent disposable baby pampers provide several advantages including convenience, comfort, outstanding leakage prevention, enhanced cleanliness, and skin care advantages. Procter & Gamble incorporates safety assurance into every step of the diaper manufacturing process with the aim of guaranteeing the security of both parents and infants. A methodical, step-by-step approach to safety evaluation begins with an in-depth analysis of new design elements and components using the general threat assessment guidelines, which may include, as necessary, controlled studies to evaluate clinical end points or an independent scientific review of safety data. Most diaper components are made of safe polymers that do not have any inherent toxicity problems. Based on their potential for skin contact, trace levels of nonpolymeric compounds, such as colorants, are evaluated. Only if new materials or design elements have been demonstrated to be secure under suggested or anticipated use, they are included in commercially available goods. Through in-market monitoring, the product's safety is maintained after introduction [32, 33].

An upper surface (nonwoven or porous film), an absorbent core, a bottom layer (plastic film, nonwoven), and a fastening system are the usual components of baby nappies, sanitary napkins, pantyliners, and adult incontinence products (Figure 7.4). An absorbent core, a cover material, and a string are the standard components of a tampon. Certain varieties of tampons also come with an applicator [34].

There are four main parts to a typical diaper. A permeable polypropylene nonwoven makes the upper covering that comes into touch with the body. The subsequent element is a layer that delivers the urine to the absorbent layer after briefly storing it there. Fluff pulp and SAP are both used to make these parts. The outer layer must maintain the diaper's integrity and stop leaks. It could have tiny air-passing microholes built into it. Polyethylene is used to create this layer. A diaper also has an elastic band, a tape-fastening method, colors, and prints. The fluff comprises the majority of

Figure 7.4: Typical structural parts of personal hygiene (baby diapers and sanitary pads).

a sanitary towel's weight (about 66% of the overall weight), as well. By weight, cotton makes up the majority of tampons (around 90% if the applicator is not included). Products made of cotton that are similar to tampons include cotton buds and cotton wadding. On the other hand, the primary ingredients of breast pads are fluff (about 78%) and SAP (about 19%) [28].

Disposable bed linen, bedding covers and draw sheets, surgeon's gowns, and diaper liners, for example, contain a number of the same substances as nappies, incontinence personal care, and sanitary towels.

The market for infant diapers on a worldwide scale includes a variety of nappies offered by producers. Cloth nappies, disposable nappies, training nappies, swim trunks, and biodegradable diapers are really among several sorts of baby diapers produced across the world. Disposable nappies can also be broken down into four different categories, including superabsorbent, normal, ultra-absorbent, and biodegradable, depending on the kind of absorbent materials being used for their production [35].

The primary drivers currently driving the growth of the baby nappies market are the rising birth rate, fast urbanization, and consistently improving economic conditions in emerging nations. Diapers are frequently used to maintain cleanliness and prevent rashes in newborns' skin. In 2020, the global market for infant diapers is anticipated to generate $59.4 billion. The demand for infant diapers appears to have been fueled by increased healthcare costs and an increasing disposable income [36].

7.3.1.2 Hydrogel-based female incontinence

Menstruation is a normal female cycle that happens every 21–28 days, lasts 5 days, and has a big impact on how well women's reproductive systems work. Menstrual periods are primarily experienced by women between the ages of 11 and 50, and they normally last 3–7 days. The average amount of blood lost during a period is 35 mL, but 10–80 mL is accepted as normal. Particularly in the beginning of the early menstrual day, the flow rate remains high, but from the second day of the cycle until the final few days, the flow rate gradually decreases. About 50–60% of the blood in menstrual fluid is mixed with uterine, cervical, and vaginal secretions as well as mucous components [37, 38].

Women in all communities in the poor world lack enough knowledge about managing menstruation hygiene, which can lead to infections, numerous illnesses that affect women's health, shame, and obstacles to employment that reduce productivity. Poor and unaware women frequently utilize filthy rags that are soiled to collect and preserve menstrual flow for their health. Additionally, impoverished women are not accustomed to wearing underwear; therefore, even if using sanitary napkins sold in stores requires wearing underwear of any kind, they are not interested in doing so. The circumstance calls for developing a type of pad that can be produced at extremely cheap cost and in an ecologically friendly manner [39].

When individuals have access to safe, inexpensive sanitary products to control their periods, it can significantly improve their overall sexual and reproductive health. The overall sexual and reproductive health may be impacted in a cascading manner, including a decline in teen pregnancy, favorable mother outcomes, and fertility. However, poor menstruation hygiene can result in major health hazards such as reproductive and urinary tract infections, which can cause infertility in the future and complicate labor and delivery. When hands are not washed after changing menstruation products, diseases including hepatitis B and thrush can spread [40, 41].

According to the research, providing sanitary napkins to girls significantly lowers the incidence of bacterial vaginosis and STDs [42]. Because of the taboos and stigma surrounding menstruation, women and girls may face discomfort, psychological stress, and even despair as a result of the absence of resources for sanitary managing of menstruation [43].

Sanitary napkins have two main purposes: they collect and hold menstrual blood and separate it from the body. No leaks, no unsightly look or color, no odor, no noise, remain in place, pleasant to wear (slim body form), and a high degree of hygiene are significant and desired qualities [44].

The first disposable item of economic relevance was the sanitary pad, which was released in 1921. Tampax Corporation unveiled the first menstruation tampon in 1933. A nonwoven rayon coversheet was used to encase cotton fiber in the first Tampax brand tampons. Nowadays, cellulose fluff is the primary absorbent material utilized in most contemporary tampons. The introduction of superabsorbents in the 1960s marked another advancement in absorbent technologies. Superabsorbent nappies were first made available in Japan in 1979, but they were not extensively used until 1984 in the West [28].

According to House et al. [47], 26% women worldwide are in the reproductive stage. The exterior and internal items, which are separated into smaller divisions, make up the two primary categories of female hygiene products that are available today:

– The category of women's exterior hygiene items includes towels and rags, washable napkins, and disposable napkins.

– Different forms of protectors are put into the vagina as internal ways of usage. They may take the blood in or keep it inside the body till the user is ready to remove it. There are two groups within this category: menstrual pads and tampons.

The market for feminine personal care items consists of a variety of products, including sanitary napkins, tampons, pantyliners, and covers. The most widely utilized feminine hygiene item is menstrual pads, and the demand from developing countries is likely to increase significantly in the coming years. Women use these items to preserve their basic hygiene. They can be made from a variety of synthetic or natural raw materials.

A biopolymer-based menstruation pad was prepared by Md. Moynul Hassan Shibly et al. They compared the product's quality to that of commercial goods. Some of the most recent sanitary pads available in the local market were analyzed as part of their research for the model's design. To improve needs including absorption, fluid retention strength, comfort, and cost reduction, six models were created by employing a variety of biopolymers such as cotton, viscose, wood pulp, sodium alginate, and carboxymethyl cellulose (CMC) in varied proportions as the core absorbent layer. The sanitized pads have undergone several tests to standardize important qualities and value, including antibacterial activity, drying capacity, and water-holding ability. The best similar result was found in a template eating sodium alginate, CMC, and cellulosic fiber, indicating that the use of sodium alginate and CMC is a potential alternative for SAP. The outer layer of sanitary napkins is made of a nonwoven fabric that has been treated with neem extract and has antimicrobial properties of greater than 90% against both gram-positive and gram-negative pathogens [45].

Similarly for use in personal hygiene, MD Nur Alam et al. prepared cellulose-based hydrogels with excellent absorption potential. They discussed a brand-new, green, aqueous-based approach for producing a new, all-natural, cellulose-based SAP (hydrogel). In this two-step procedure, cellulose was first converted to CMC by reaction with sodium monochloroacetate, and then cross-linked with epichlorohydrin. The newly created hydrogels' water-holding value (WHV) in deionized water (d-water) was 725 g d-water/g gel, which is considerably higher than the WHV of any other commercially available superabsorbent cellulose-based material (WHV of 10–100 g/g) and roughly equivalent to the WHV of commercially produced (polyacrylate) SAP gels (up to 1,000 g/g). The aforementioned freshly prepared hydrogels could be ready to compete with synthetic SAPs in innovative applications including food, medicine, and personal care products [46].

7.4 Conclusion

Hydrogels have been around for more than 50 years, yet none can deny their allure. Due to their unique characteristics, hydrogels have captured the interest of both the scientific and business sectors. However, despite the fact that research on their potential has been widely disseminated, not all of the formulations that have been prepared have been commercially successful. Only considering biomedical uses, hydrogels have been used in a variety of ways and concentrations in a number of industries, including drug delivery, wound healing, and personal hygiene products. The researcher must pay attention to design and synthesize the biocompatable and biodegradable hydrogels both for the new infant and women personal care products. The newly synthesize hydrogels are in high demand to cope with the medical and environmental issues.

References

[1] Cascone, S. & Lamberti, G. (2020). Hydrogel-based commercial products for biomedical applications: A review. Int J pharm, 573, 118803.

[2] Ahmed, E. M. (2015). Hydrogel: Preparation, characterization, and applications: A review. J Adv Res, 6(2), 105–121.

[3] Kopecek, J. (2009). Hydrogels: From soft contact lenses and implants to self-assembled nanomaterials. J Polym Sci Part A, 47(22), 5929–5946.

[4] Ahmad, Z., et al. (2022). Versatility of hydrogels: From synthetic strategies, classification, and properties to biomedical applications. Gels, 8(3), 167.

[5] Sharma, S., Parmar, A., & Mehta, S. (2018). Hydrogels: From simple networks to smart materials – advances and applications. J Drug Target, 627–672.

[6] Buwalda, S. J., et al. (2014). Hydrogels in a historical perspective: From simple networks to smart materials. J Control Release, 190, 254–273.

[7] Aswathy, S., Narendrakumar, U., & Manjubala, I. (2020). Commercial hydrogels for biomedical applications. Heliyon, 6(4), e03719.

[8] Ratner, B. D. & Hoffman, A. S. (1976). Synthetic Hydrogels for Biomedical Applications. ACS Publications.

[9] Seliktar, D. (2012). Designing cell-compatible hydrogels for biomedical applications. Science, 336(6085), 1124–1128.

[10] Chawla, P., et al. (2014). Hydrogels: A journey from diapers to gene delivery. Mini Rev Med Chem, 14(2), 154–167.

[11] Nizam El-Din, H. M. (2012). Surface coating on cotton fabrics of new multilayer formulations based on superabsorbent hydrogels synthesized by gamma radiation designed for diapers. J Appl Polym Sci, 125(S2), E180–E186.

[12] Catoira, M. C., et al. (2019). Overview of natural hydrogels for regenerative medicine applications. J Mater Sci Mater Med, 30(10), 1–10.

[13] Madduma-Bandarage, U. S. & Madihally, S. V. (2021). Synthetic hydrogels: Synthesis, novel trends, and applications. J Appl Polym Sci, 138(19), 50376.

[14] Palmese, L. L., et al. (2019). Hybrid hydrogels for biomedical applications. Curr Opin Chem Eng, 24, 143–157.

[15] Leal, D., et al. (2013). Preparation and characterization of hydrogels based on homopolymeric fractions of sodium alginate and PNIPAAm. Carbohydr Polym, 92(1), 157–166.

[16] Taşdelen, B., et al. (2004). Preparation of poly (N-isopropylacrylamide/itaconic acid) copolymeric hydrogels and their drug release behavior. Int J pharm, 278(2), 343–351.

[17] Silverstein, M. S. (2020). Interpenetrating polymer networks: So happy together?. Polymer, 207, 122929.

[18] Hu, Z. & Huang, G. (2003). A new route to crystalline hydrogels, guided by a phase diagram. Angew Chem, 115(39), 4947–4950.

[19] Kurt, B., et al. (2016). High-strength semi-crystalline hydrogels with self-healing and shape memory functions. Eur Polym J, 81, 12–23.

[20] Agren, M. (1998). An amorphous hydrogel enhances epithelialisation of wounds. Acta Dermatol Venereologica, 78(2), 119–122.

[21] Tong, R., et al. (2019). Highly stretchable and compressible cellulose ionic hydrogels for flexible strain sensors. Biomacromolecules, 20(5), 2096–2104.

[22] Rivero, R., et al. (2017). Physicochemical properties of ionic and non-ionic biocompatible hydrogels in water and cell culture conditions: Relation with type of morphologies of bovine fetal fibroblasts in contact with the surfaces. Colloids Surf B Biointerfaces, 158, 488–497.

[23] Akhtar, M. F., Hanif, M., & Ranjha, N. M. (2016). Methods of synthesis of hydrogels . . . A review. Saudi Pharmaceut J, 24(5), 554–559.

[24] Hu, W., et al. (2019). Advances in crosslinking strategies of biomedical hydrogels. Biomater Sci, 7(3), 843–855.

[25] Albinger, N., Hartmann, J., & Ullrich, E. (2021). Current status and perspective of CAR-T and CAR-NK cell therapy trials in Germany. Gene Ther, 28(9), 513–527.

[26] Ahmad, Z., et al. (2021). Bacterial cellulose composites, synthetic strategies, and applications. In Bacterial Cellulose (pp. 201–211). CRC Press.

[27] Barman, A., Katkar, P. M., & Asagekar, S. D. (2018). Natural and sustainable raw materials for sanitary napkin. Man-Made Textil India, 46, 12.

[28] Bashari, A., Rouhani Shirvan, A., & Shakeri, M. (2018). Cellulose-based hydrogels for personal care products. Polym Adv Technol, 29(12), 2853–2867.

[29] Mistry, P. A., et al. (2023). Chitosan superabsorbent biopolymers in sanitary and hygiene applications. Int J Polym Sci, 2023.

[30] Kamat, M. & Malkani, R. (2003). Disposable diapers: A hygienic alternative. Indian J Pediatr, 70(11), 879–881.

[31] Zainal, S. H., et al. (2021). Preparation of cellulose-based hydrogel: A review. J Mater Res Technol, 10, 935–952.

[32] Kosemund, K., et al. (2009). Safety evaluation of superabsorbent baby diapers. Regul Toxicol Pharmacol, 53(2), 81–89.

[33] Płotka-Wasylka, J., et al. (2022). End-of-life management of single-use baby diapers: Analysis of technical, health and environment aspects. Sci Total Environ, 155339.

[34] Kakonke, G., et al. (2019). Review on the manufacturing and properties of nonwoven superabsorbent core fabrics used in disposable diapers.

[35] Edana. (2008). Sustainability Report 2007–2008: Absorbent Hygiene Products. Edana Brussels.

[36] Brochmann, N. & Dahl, E. S. (2018). The Wonder Down Under: The Insider's Guide to the Anatomy, Biology, and Reality of the Vagina. Quercus.

[37] Critchley, H. O., et al. (2020). Menstruation: Science and society. Am J Clin Exp Obstet Gynecol, 223(5), 624–664.

[38] Jabbour, H. N., et al. (2006). Endocrine regulation of menstruation. Endocrine Rev, 27(1), 17–46.

[39] Montgomery, P., et al. (2016). Menstruation and the cycle of poverty: A cluster quasi-randomised control trial of sanitary pad and puberty education provision in Uganda. Plos One, 11(12), e0166122.

[40] Dasgupta, A. & Sarkar, M. (2008). Menstrual hygiene: How hygienic is the adolescent girl?. Indian J Community Med, 33(2), 77.

[41] Das, P., et al. (2015). Menstrual hygiene practices, WASH access and the risk of urogenital infection in women from Odisha, India. PloS One, 10(6), e0130777.

[42] Benshaul-Tolonen, A., et al. (2020). Measuring menstruation-related absenteeism among adolescents in low-income countries. The Palgrave handbook of critical menstruation studies, 705–723.

[43] Sweetman, C. & Medland, L. (2017). Introduction: Gender and water, sanitation and hygiene. Gender Dev, 25(2), 153–166.

[44] Woo, J., et al. (2019). Systematic review on sanitary pads and female health. Ewha Med J, 42(3), 25–38.

[45] Shibly, M., et al. (2021). Development of biopolymer-based menstrual pad and quality analysis against commercial merchandise. Bull Natl Res Cent, 45(1), 1–13.

[46] Alam, M. N., Islam, M. S., & Christopher, L. P. (2019). Sustainable production of cellulose-based hydrogels with superb absorbing potential in physiological saline. ACS Omega, 4(5), 9419–9426.

[47] House, S., Mahon, T., & Cavill, S. (2013). Menstrual hygiene matters: a resource for improving menstrual hygiene around the world. Reprod Health Matt, 21(41), 257–259.

Shahid Ali Khan, Zubair Ahmad, Zhiduan Cai, Guibin Xu*

Chapter 8
Hydrogels for kidney carcinoma

Abstract: In this chapter, the role of hydrogels is discussed in kidney carcinoma. These versatile biomaterials, known for their unique properties and diverse applications in biomedicine, extend their influence far beyond drug delivery. From tissue engineering to wound healing and more, hydrogels have become pivotal players in addressing the multifaceted challenge posed by kidney carcinoma. As the field of hydrogel technology continues to advance, it holds immense promise in revolutionizing personalized medicine. By providing effective, patient-centric solutions, hydrogels emerge as powerful allies in the fight against this formidable disease. The integration of hydrogels into kidney carcinoma therapy represents a promising frontier. These biomaterials offer the distinct advantages of localized drug delivery, biocompatibility, and adaptability to the tumor microenvironment. Researchers' tireless efforts in refining drug-eluting hydrogels and exploring innovative approaches provide a beacon of hope for improved treatment outcomes and enhanced patient quality of life.

Keywords: Hydrogels, kidney carcinoma, biomedical applications

Acknowledgments: The authors highly acknowledge the Department of Urology, the Fifth Affiliated Hospital of Guangzhou Medical University China; Department of Chemistry, School of Natural Sciences, National University of Science and Technology Pakistan; and Department of Chemistry, University of Swabi Pakistan, for their collaborating work.

*Corresponding author: Guibin Xu, Department of Urology, Key Laboratory of Biological Targeting Diagnosis, Therapy and Rehabilitation of Guangdong Higher Education Institutes, The Fifth Affiliated Hospital of Guangzhou Medical University, Guangzhou Medical University, Guangzhou 510700, China, e-mail: uro_xgb@163.com
Shahid Ali Khan, Department of Urology, Key Laboratory of Biological Targeting Diagnosis, Therapy and Rehabilitation of Guangdong Higher Education Institutes, The Fifth Affiliated Hospital of Guangzhou Medical University, Guangzhou Medical University, Guangzhou 510700, China; Department of Chemistry, School of Natural Sciences, National University of Science and Technology (NUST), Islamabad 44000, Pakistan
Zubair Ahmad, Department of Chemistry, University of Swabi, Khyber Pakhtunkhwa, Anbar 23561, Pakistan
Zhiduan Cai, Department of Urology, Key Laboratory of Biological Targeting Diagnosis, Therapy and Rehabilitation of Guangdong Higher Education Institutes, The Fifth Affiliated Hospital of Guangzhou Medical University, Guangzhou Medical University, Guangzhou 510700, China

https://doi.org/10.1515/9783111334080-008

8.1 Introduction to kidney carcinoma

Kidney carcinoma, also known as renal cell carcinoma (RCC), stands as a significant medical concern characterized by the malignant transformation of cells within the kidneys. Representing a substantial portion of all kidney cancer cases, RCC accounts for approximately 90% of diagnoses, establishing itself as one of the most prevalent malignancies worldwide. This cancer transcends gender lines, affecting both men and women and frequently presents itself during adulthood, with a pronounced incidence in individuals aged 50–70 years [1, 2].

8.1.1 Epidemiology

The incidence rate of RCC accounts for 3–5% of adult malignancies [3, 4], ranking third among male genitourinary malignancies after prostate cancer and bladder cancer. There are significant differences in the incidence rates of RCC among countries and regions. In most countries and regions, the incidence rate of RCC is increasing year by year, but in developed countries, the mortality rate tends to stabilize or decrease. In the United States, kidney cancer is one of the top 10 most common cancers, with approximately 81,800 new cases and 14,890 deaths reported in 2023. The male-to-female ratio of incidence is about 1.8:1 [3]. In China, the incidence rate of RCC is declining, but the risk for urban populations is still 2.34 times higher than that for rural areas, and the risk for males is still higher than for females with a ratio of about 1.7:1 [4].

8.1.2 Etiology

The etiology of RCC is still unclear, but it is associated with genetics, smoking, alcohol consumption, obesity, hypertension and antihypertensive drugs, and diabetes. Smoking, obesity, and hypertension are currently recognized as risk factors for RCC, with 60% of patients being related to these three factors. Therefore, quitting smoking and controlling weight and blood pressure are important measures for preventing the occurrence of RCC [5, 6]. Environmental risk factors such as water pollution (such as arsenic contamination), exposure to industrial solvents, or occupational exposure (such as trichloroethylene, herbicides, pesticides, asbestos, and copper sulfate) have been reported to be associated with the development of RCC and increased mortality rate [7]. Most cases of RCC are sporadic nonhereditary cases while 5% have a familial hereditary component [7]. Currently, there are 10 hereditary cancer susceptibility syndromes associated with genetic risk for kidney cancer identified by 12 genes, including BAP1, FLCN, Translocation chr 3, FH, MET, PTEN, SDHB, SDHC, SDHD, TSC1, TSC2, and VHL [8].

8.1.3 Diagnosis

The clinical diagnosis of RCC mainly relies on imaging examinations, combined with clinical manifestations and laboratory tests to determine the clinical stage. Over 50% of RCC patients are asymptomatic and are usually diagnosed incidentally through abnormal findings in imaging examinations [9]. The typical triad of renal cancer, consisting of gross hematuria, flank pain, and palpable abdominal mass, is now uncommon. Hematuria is the most common symptom, and the presence of the triad usually indicates advanced-stage tumors [10, 11]. Approximately 20% of patients have paraneoplastic syndromes such as hypertension, hypercalcemia, hyperglycemia, polycythemia, and coagulation abnormalities [9]. Some characteristic manifestations may also occur; for example, coughing or hemoptysis may indicate lung metastasis; bone pain or fractures may suggest bone metastasis; headaches may be a sign of brain metastasis; and cervical lymphadenopathy can indicate lymph node metastasis. Patients with inferior vena cava tumor thrombus may experience lower limb edema. If left renal vein thrombosis is present along with inferior vena cava tumor thrombus, there may be varicocele on the left side. Pathological examination is necessary for confirming the diagnosis of RCC by determining the extent of invasion based on postoperative histology staging. Definitive diagnosis requires pathological examination. According to histopathological diagnosis, 90% of kidney cancers are RCCs, which mainly include clear cell carcinoma (70–75%), papillary RCC, and chromophobe RCC [12].

8.1.4 Treatment

For any localized RCC, the preferred treatment method is surgical resection, including partial nephrectomy with preservation of the kidney unit and radical nephrectomy. In addition, for patients with localized RCC who are not suitable for surgery, options such as thermal ablation, cryoablation, and radiofrequency ablation can be considered [13]. Radical nephrectomy is the standard treatment for localized primary RCC; however, approximately 25% of these patients will experience distant metastasis after surgery. Furthermore, about 30% of primary RCC patients have local progression or distant metastasis at the time of diagnosis [14]. The treatment of metastatic RCC is more complex and challenging, primarily focusing on systemic drug therapy supplemented by palliative surgery or radiation therapy.

The systemic treatment for metastatic RCC includes chemotherapy, radiotherapy, targeted therapy, and immunotherapy. The efficacy of chemotherapy is limited and often combined with experimental immunotherapy. Radiotherapy is mainly used for distant metastasis or local tumor recurrence to relieve pain and improve quality of life. Molecular targeted drugs and immune checkpoint inhibitors are currently the main treatment options. The first-line drugs for molecular targeted therapy include sorafenib, sunitinib, pazopanib, cabozantinib, and lenvatinib [15–19], while the second-line drugs include axitinib, everolimus, and temsirolimus [20–22]. Immune checkpoint inhibitors are mainly used in combination with targeted therapy or other immune agents. Common regimens include pembrolizumab plus axitinib, nivolumab plus ipilimumab, avelumab plus axitinib, pembrolizumab plus lenvatinib, and nivolumab plus cabozantinib [23–27]. After the failure of first-line treatment, alternative options such as lenvatinib plus everolimus, lenvatinib plus pembrolizumab, ipilimumab plus nivolumab, and vorolanib plus everolimus can be considered [28–31].

8.1.5 Prognosis

The most important indicator for the prognosis of RCC is pathological staging, with a 5-year survival rate of 85–90% for patients diagnosed with stage I or II cancer [13]. Univariate analysis shows that the prognosis of RCC is related to histological subtypes, with chromophobe and papillary RCCs having better prognoses than clear cell carcinoma, with 5-year survival rates of 88%, 91%, and 71%, respectively [32–35]. Within the papillary subtype, type I tumors are low grade and have a better prognosis, while type II tumors are high grade and more likely to metastasize, resulting in a poorer prognosis [36]. However, comprehensive analysis considering multiple factors such as tumor grading and staging reveals that histological subtypes cannot be considered as independent prognostic factors [33]. Currently, available molecular markers lack accuracy in predicting the prognosis of RCC and require further research validation before being recommended for clinical use [37–39]. There are various prognostic evaluation systems for predicting the prognosis of RCC. The commonly used systems include UISS (UCLA Integrated Staging System) and SSIGN (Stage Size Grade Necrosis) for localized and locally advanced RCCs; IMDC (International Metastatic Renal Cell Carcinoma Database Consortium) and MSKCC (Memorial Sloan-Kettering Cancer Center) scoring system for advanced/metastatic RCCs to assess risk levels [40–44].

The next section provides a brief discussion of hydrogels followed by some clinical application and their role in kidney carcinoma.

8.2 Brief introduction of hydrogels

Wichterle and Lim introduced the term "hydrogels" for the first time in 1960 and later it became the most emerging biomaterial on a large scale [45, 46]. Owing to its hydrophilic and three-dimensional nature, hydrogel absorbs and retains water more than its weight. Despite the large amount of adsorbed water, it is not soluble in water. A literature survey revealed that the capacity of water absorption is higher in synthetic polymeric hydrogels compared to their natural polymer hydrogel counterparts. One can increase the hydrophilic and water-retaining capability of polymeric hydrogels by introducing some functional groups; for instance, $-OH$, SO_3H, and NH_2 [47]. These functional groups contribute to the network's hydrophilicity. Owing to its high water-retaining ability and soft nature, a hydrogel is like a natural tissue [48]; thus, hydrogel is one of the most suitable candidates for biomedical applications [49]. Hydrogels are stimuli responsive and can change their behavior with external stimuli, for instance, a change in temperature, pH, ions, light, or chemical or biological change [50]. Moreover, the change in hydrogel shapes depends on the degree of external stimuli [51–54]. The extent and nature of these responses depend on the specific environmental changes and the design of the hydrogel, enabling precise control over their behavior in different applications. This responsiveness has led to the development of drug-eluting hydrogels, which have shown great promise in the treatment of various diseases, including kidney carcinoma [55]. Kidney carcinoma, particularly RCC, is a formidable challenge in oncology due to its resistance to conventional therapies and limited treatment options [56]. However, drug-eluting hydrogels have emerged as a potential breakthrough in the management of renal carcinoma [57]. Drug-eluting hydrogels are designed to deliver therapeutic agents, such as chemotherapeutic drugs or targeted therapies, directly to the tumor site. These hydrogels can be administered locally, providing sustained drug release over an extended period. In the case of kidney carcinoma, localized drug delivery is especially advantageous as it minimizes systemic exposure, reducing the risk of systemic side effects associated with conventional chemotherapy [58]. The choice of drug and hydrogel composition is critical in the development of drug-eluting hydrogels for kidney carcinoma [59]. Researchers are exploring various strategies to optimize drug release kinetics, ensuring that the therapeutic agent is released at a rate that effectively targets the tumor while minimizing damage to the healthy surrounding tissue [60]. Additionally, the stimuli-responsive nature of hydrogels allows for controlled drug release triggered by factors specific to the tumor microenvironment, such as pH or enzyme levels [61]. Furthermore, drug-eluting hydrogels can be used in combination with other treatment modalities, such as surgery or radiation therapy, to enhance therapeutic outcomes. These multimodal approaches hold great promise for improving the prognosis of patients with kidney carcinoma [62].

So hydrogels, with their exceptional properties and responsiveness to external stimuli, have opened up new avenues in the development of drug delivery systems for the treatment of kidney carcinoma. The ability to precisely control drug release at the tumor site, reduce systemic side effects, and combine treatments for synergistic effects makes drug-eluting hydrogels a promising tool to fight against this challenging cancer. As research to advance, we can anticipate even more innovative approaches to improve the effectiveness of drug-eluting hydrogels in the treatment of kidney carcinoma and other related diseases [58].

8.3 Hydrogels in the clinic

Hydrogels have made significant strides in clinical applications due to their unique properties, which make them versatile biomaterials [63, 64]; for instance, owing to their biocompatible nature, it can be well tolerated by the human body and can be used in the tissue mimicry. Their soft, water-rich structure closely mimics the natural environment of tissues, reducing the risk of inflammation or adverse reactions when used in clinical applications [65]. It can be extensively used in wound healing for several reasons, because it creates a moist environment for wound which is essential for optimal healing by promoting angiogenesis, cell migration, and tissue regeneration. Moreover, the moist environment can relieve pain at the site of wound healing by providing a cooling effect. Additionally, some hydrogels can absorb excess wound exudate, keeping the wound clean and reducing the risk of infection. Hydrogels can also be formulated to provide localized pain relief or release antimicrobial agents and growth factors, further aiding in the healing process [66]. Many reports are available on hydrogels in the drug delivery system, for instance, it is reported to deliver medications, growth factors, or other therapeutic agents directly to the affected tissue. This localized drug delivery enhances the treatment efficacy while minimizing side effects [67]. Hydrogels serve as valuable scaffolds for tissue engineering. When parts of the kidney need repair or replacement following tumor resection, hydrogel-based scaffolds can support the growth of new tissue, offering promise for kidney regeneration [68]. Hydrogels also find applications in diagnostics and imaging. Their ability to retain contrast agents or nanoparticles makes them valuable tools for enhancing tumor visibility or assessing kidney function through imaging techniques such as magnetic resonance imaging or ultrasound [69, 70]. Hydrogels play a role in immunotherapy by acting as carriers for immune checkpoint inhibitors and other immunomodulatory agents. Immunotherapy has emerged as a promising approach for the treatment of advanced kidney carcinoma [71]. Hydrogel-based drug delivery systems can be employed to administer immunotherapeutic agents, enhancing the patient's immune response against cancer cells [72].

8.4 Hydrogels for kidney therapy

In the realm of kidney therapy, hydrogels have emerged as versatile and innovative tools with the potential to revolutionize treatment approaches. The in-depth applications of hydrogels in kidney therapy, emphasizing their capacity for precise drug delivery, biocompatibility, responsiveness to external stimuli, and recent advancements make them suitable for kidney-related diseases [73, 74].

8.4.1 Role of hydrogels in facilitating drug delivery to the kidneys

Hydrogels have garnered significant attention for their pivotal role in facilitating drug delivery to the kidneys. These versatile materials have shown their effectiveness in achieving local and controlled release of desired drugs, ensuring that drugs reach the renal tissue precisely where they are needed. Additionally, hydrogels have shown promise in protecting and enhancing the effects of concurrently transplanted proregenerative cells. By encapsulating drugs within a hydrogel matrix, scientists can fine-tune the release kinetics, optimizing therapeutic outcomes while minimizing adverse effects on other organs. This innovative approach holds great promise for the treatment of kidney diseases, offering a tailored and targeted solution to improve patient care in the field of nephrology [60]. The different features of drug delivery via hydrogels are discussed below.

8.4.1.1 Tailored drug delivery

Hydrogels can be precisely tailored to deliver therapeutic agents to the kidneys, offering a targeted and controlled approach. This is particularly advantageous in kidney therapy, as it allows for the delivery of drugs directly to the affected renal tissue while minimizing systemic exposure. By encapsulating medications within hydrogels, the release can be controlled, ensuring a sustained and localized drug presence in the kidneys [55].

8.4.1.2 Reducing systemic side effects

The ability to target drug delivery to the kidneys holds immense potential for reducing systemic side effects. Kidney therapy often involves medications that can have adverse effects on other organs. By using hydrogels as drug carriers, therapeutic compounds can be confined to the kidney region, limiting collateral damage to healthy tissues and organs [75].

8.4.1.3 Biocompatibility and personalized treatment

Hydrogels are inherently biocompatible, making them well-suited for applications within the intricate environment of the kidneys. Their water content and softness closely mimic natural tissues, reducing the likelihood of adverse reactions. This biocompatibility ensures that hydrogels can seamlessly integrate with renal tissues, minimizing the risk of inflammation or immune responses [76]. The versatility of hydrogels allows for personalized treatment approaches in kidney therapy. They can be engineered to match the specific needs of individual patients, incorporating factors like drug type, dosage, and release kinetics. This personalization is especially valuable in managing kidney diseases, where treatment responses can vary widely among patients [76].

8.4.2 Recent advancements in drug-eluting hydrogels

Recent advancements have seen the development of drug-eluting hydrogels specifically designed for kidney therapy. These hydrogels are engineered to release medications in response to specific triggers or stimuli, ensuring optimal drug delivery at the right time and place. This targeted approach enhances treatment efficacy while minimizing side effects [77].

8.4.2.1 Principle of drug elution

Drug-eluting hydrogels are formulated to encapsulate drugs or therapeutic compounds. Through diffusion or degradation, these hydrogels gradually release the drugs over time, ensuring a sustained therapeutic effect. The principle of drug elution in drug-eluting hydrogels is based on a finely tuned balance between the hydrogel's structure and the properties of the encapsulated drugs. These hydrogels are engineered to control the release of therapeutic agents through diffusion or gradual degradation of the hydrogel matrix. This controlled release mechanism is essential for maintaining a steady drug concentration at the tumor site over an extended period, optimizing the effectiveness of the treatment [78].

8.4.2.2 Precise dosage control

The advantage of drug-eluting hydrogels lies in their ability to provide precise dosage control. This is particularly crucial in kidney carcinoma treatment, where specific drug concentrations are required for optimal results. Kidney carcinoma treatment often demands precise control over drug dosages to achieve therapeutic efficacy while minimizing side effects. Drug-eluting hydrogels excel in this aspect by allowing for meticulous dosage control. Researchers can fine-tune the hydrogel's composition and structure to achieve the desired release rate, ensuring that the drugs are delivered at the right concentration to effectively target cancer cells while sparing healthy tissues [79].

8.4.2.3 Minimized systemic effects

By delivering drugs directly to the tumor site, drug-eluting hydrogels minimize systemic side effects. This localized approach enhances the therapeutic index of the drugs while reducing harm to healthy tissues. The localized drug delivery achieved with drug-eluting hydrogels significantly reduces the risk of systemic side effects commonly associated with traditional systemic drug administration. By directly targeting the tumor site, these hydrogels ensure that the therapeutic agents exert their effects primarily where needed, thus sparing healthy tissues from unnecessary exposure. This approach not only enhances the overall safety of the treatment but also contributes to a higher quality of life for patients [80].

8.4.2.4 Tailored formulations

Scientists can customize drug-eluting hydrogels to accommodate a wide range of therapeutic agents, including chemotherapy drugs, targeted therapies, and immunomodulators. This versatility allows for tailored treatment strategies. The versatility of drug-eluting hydrogels extends to their ability to incorporate various therapeutic agents. Researchers can tailor these hydrogels to deliver chemotherapy drugs, targeted therapies that specifically attack cancer cells, or immunomodulatory agents that harness the patient's immune system against the tumor. This flexibility enables oncologists to design personalized treatment strategies that align with the unique characteristics of each patient's kidney carcinoma [81].

8.4.2.5 Nanotechnology and hydrogels

Nanotechnology has been integrated with hydrogels to create nanogels, which are miniature hydrogel particles. These nanogels can carry a variety of therapeutic agents, including small molecules, proteins, and nucleic acids. Researchers are exploring their potential for precise drug delivery to the kidneys, opening new avenues for kidney therapy [60].

8.4.2.6 Renal replacement technologies

Hydrogels are being explored as key components in the development of artificial kidneys and renal replacement therapies. These advanced systems aim to mimic the functions of natural kidneys and provide alternatives for patients with end-stage renal disease. Hydrogels play a critical role in creating functional, biocompatible membranes and filtration systems for these devices [74].

8.4.2.7 Combination therapies

Hydrogels enable combination therapy by incorporating multiple therapeutic agents. In the case of kidney carcinoma, this approach can enhance treatment efficacy and address the heterogeneity of the disease, improving outcomes for patients [82].

8.4.2.8 Research and development

Ongoing research efforts are focused on optimizing the design of drug-eluting hydrogels for kidney carcinoma therapy. This includes exploring novel drug combinations, improving drug release kinetics, and enhancing the overall effectiveness of the hydrogel delivery system. The field of drug-eluting hydrogels for kidney carcinoma therapy is continually evolving. Researchers are dedicated to advancing the technology by investigating novel drug combinations that exhibit synergistic effects, fine-tuning drug release kinetics to match specific treatment requirements, and optimizing the overall performance of the hydrogel delivery system. These ongoing efforts aim to further improve treatment outcomes, reduce side effects, and expand the application of drug-eluting hydrogels in the fight against kidney carcinoma [78].

8.5 Conclusion

The field of hydrogels presents an exciting frontier in the realm of biomedical applications, particularly in the context of addressing the challenges posed by kidney carcinoma. Hydrogels, with their unique properties and versatility, offer a wide range of applications in the field of biomedicine. This chapter has highlighted the pivotal role of hydrogels in various aspects of healthcare, from tissue engineering to wound healing and specially focusing on renal carcinoma. One of the most promising aspects of hydrogel technology is its potential to revolutionize the personalized medicine. By harnessing the capabilities of hydrogels, researchers and healthcare professionals can provide effective and patient-centric solutions for individuals battling kidney carcinoma. These biomaterials offer distinct advantages, including localized drug delivery, biocompatibility, and adaptability to the complex tumor microenvironment. The tireless efforts of researchers in developing drug-eluting hydrogels and exploring innovative approaches provide a beacon of hope for improved treatment outcomes and enhanced patient quality of life. As the field of hydrogel technology continues to advance, it holds immense promise in the fight against this formidable disease. Hydrogels are emerging as powerful allies, offering new avenues for more targeted and efficient therapies in the battle against kidney carcinoma.

References

[1]   Dahle, D. O., et al. (2022). Renal cell carcinoma and kidney transplantation: A narrative review. Transplantation, 106, 1.

[2]   Dell'Atti, L., Bianchi, N., & Aguiari, G. (2022). New therapeutic interventions for kidney carcinoma: Looking to the future. Cancers, 14(15), 3616.

[3]   Siegel, R. L., et al. (2023). Cancer statistics. CA Cancer J Clin, 73(1), 17–48.

[4]   Xia, C., et al. (2022). Cancer statistics in China and United States, 2022: Profiles, trends, and determinants. Chin Med J (Engl), 135(5), 584–590.

[5]   Scelo, G. & Larose, T. L. (2018). Epidemiology and risk factors for kidney cancer. J Clin Oncol, 36(36), JCO2018791905.

[6]   Chow, W. H., Dong, L. M., & Devesa, S. S. (2010). Epidemiology and risk factors for kidney cancer. Nat Rev Urol, 7(5), 245–257.

[7]   Webster, B. R., Gopal, N., & Ball, M. W. (2022). Tumorigenesis mechanisms found in hereditary renal cell carcinoma: A review. Genes (Basel), 13(11).

[8]   Haas, N. B. & Nathanson, K. L. (2014). Hereditary kidney cancer syndromes. Adv Chronic Kidney Dis, 21(1), 81–90.

[9]   Campbell, S. C., et al. (2009). Guideline for management of the clinical T1 renal mass. J Urol, 182(4), 1271–1279.

[10]   Cohen, H. T. & McGovern, F. J. (2005). Renal-cell carcinoma. N Engl J Med, 353(23), 2477–2490.

[11]   Loo, R. K., et al. (2013). Stratifying risk of urinary tract malignant tumors in patients with asymptomatic microscopic hematuria. Mayo Clin Proc, 88(2), 129–138.

[12]   Yang, J., Wang, K., & Yang, Z. (2023). Treatment strategies for clear cell renal cell carcinoma: Past, present and future. Front Oncol, 13, 1133832.

[13]   Campbell, S. C., et al. (2021). Renal mass and localized renal cancer: Evaluation, management, and follow-up: AUA guideline: Part I. J Urol, 206(2), 199–208.

[14]   Choueiri, T. K. & Motzer, R. J. (2017). Systemic therapy for metastatic renal-cell carcinoma. N Engl J Med, 376(4), 354–366.

[15]   Escudier, B., et al. (2009). Sorafenib for treatment of renal cell carcinoma: Final efficacy and safety results of the phase III treatment approaches in renal cancer global evaluation trial. J Clin Oncol, 27(20), 3312–3318.

[16]   Motzer, R. J., et al. (2009). Overall survival and updated results for sunitinib compared with interferon alfa in patients with metastatic renal cell carcinoma. J Clin Oncol, 27(22), 3584–3590.

[17]   Sternberg, C. N., et al. (2010). Pazopanib in locally advanced or metastatic renal cell carcinoma: Results of a randomized phase III trial. J Clin Oncol, 28(6), 1061–1068.

[18]   Zhou, A. P., et al. (2019). Anlotinib versus sunitinib as first-line treatment for metastatic renal cell carcinoma: A randomized phase II clinical trial. Oncologist, 24(8), e702–e708.

[19]   Choueiri, T. K., et al. (2018). Cabozantinib versus sunitinib as initial therapy for metastatic renal cell carcinoma of intermediate or poor risk (Alliance A031203 CABOSUN randomised trial): Progression-free survival by independent review and overall survival update. Eur J Cancer, 94, 115–125.

[20]   Rini, B. I., et al. (2011). Comparative effectiveness of axitinib versus sorafenib in advanced renal cell carcinoma (AXIS): A randomised phase 3 trial. Lancet, 378(9807), 1931–1939.

[21]   Motzer, R. J., et al. (2010). Phase 3 trial of everolimus for metastatic renal cell carcinoma: Final results and analysis of prognostic factors. Cancer, 116(18), 4256–4265.

[22]   Rini, B. I., et al. (2020). Tivozanib versus sorafenib in patients with advanced renal cell carcinoma (TIVO-3): A phase 3, multicentre, randomised, controlled, open-label study. Lancet Oncol, 21(1), 95–104.

[23] Powles, T., et al. (2020). Pembrolizumab plus axitinib versus sunitinib monotherapy as first-line treatment of advanced renal cell carcinoma (KEYNOTE-426): Extended follow-up from a randomised, open-label, phase 3 trial. Lancet Oncol, 21(12), 1563–1573.

[24] Motzer, R. J., et al. (2018). Nivolumab plus ipilimumab versus sunitinib in advanced renal-cell carcinoma. N Engl J Med, 378(14), 1277–1290.

[25] Motzer, R. J., et al. (2019). Avelumab plus axitinib versus sunitinib for advanced renal-cell carcinoma. N Engl J Med, 380(12), 1103–1115.

[26] Motzer, R., et al. (2021). Lenvatinib plus pembrolizumab or everolimus for advanced renal cell carcinoma. N Engl J Med, 384(14), 1289–1300.

[27] Choueiri, T. K., et al. (2020). 696O_PR Nivolumab + cabozantinib vs sunitinib in first-line treatment for advanced renal cell carcinoma: First results from the randomized phase III CheckMate 9ER trial. Ann Oncol, 31.

[28] Motzer, R. J., et al. (2015). Lenvatinib, everolimus, and the combination in patients with metastatic renal cell carcinoma: A randomised, phase 2, open-label, multicentre trial. Lancet Oncol, 16(15), 1473–1482.

[29] Lee, C. H., et al. (2021). Lenvatinib plus pembrolizumab in patients with either treatment-naive or previously treated metastatic renal cell carcinoma (Study 111/KEYNOTE-146): A phase 1b/2 study. Lancet Oncol, 22(7), 946–958.

[30] Hammers, H. J., et al. (2017). Safety and efficacy of nivolumab in combination with ipilimumab in metastatic renal cell carcinoma: The CHECKMATE 016 study. J Clin Oncol, 35(34), 3851–3858.

[31] Sheng, X., et al. (2020). Phase 1 trial of vorolanib (CM082) in combination with everolimus in patients with advanced clear-cell renal cell carcinoma. EBioMedicine, 55, 102755.

[32] Cheville, J. C., et al. (2003). Comparisons of outcome and prognostic features among histologic subtypes of renal cell carcinoma. Am J Surg Pathol, 27(5), 612–624.

[33] Patard, J. J., et al. (2005). Prognostic value of histologic subtypes in renal cell carcinoma: A multicenter experience. J Clin Oncol, 23(12), 2763–2771.

[34] Wahlgren, T., et al. (2013). Treatment and overall survival in renal cell carcinoma: A Swedish population-based study (2000–2008). Br J Cancer, 108(7), 1541–1549.

[35] Li, P., et al. (2016). Survival among patients with advanced renal cell carcinoma in the pretargeted versus targeted therapy eras. Cancer Med, 5(2), 169–181.

[36] Delahunt, B., et al. (2001). Morphologic typing of papillary renal cell carcinoma: Comparison of growth kinetics and patient survival in 66 cases. Hum Pathol, 32(6), 590–595.

[37] Sim, S. H., et al. (2012). Prognostic utility of pre-operative circulating osteopontin, carbonic anhydrase IX and CRP in renal cell carcinoma. Br J Cancer, 107(7), 1131–1137.

[38] Li, G., et al. (2008). Serum carbonic anhydrase 9 level is associated with postoperative recurrence of conventional renal cell cancer. J Urol, 180(2), 510–513.

[39] Choueiri, T. K., et al. (2014). A phase I study of cabozantinib (XL184) in patients with renal cell cancer. Ann Oncol, 25(8), 1603–1608.

[40] Zisman, A., et al. (2001). Improved prognostication of renal cell carcinoma using an integrated staging system. J Clin Oncol, 19(6), 1649–1657.

[41] Frank, I., et al. (2002). An outcome prediction model for patients with clear cell renal cell carcinoma treated with radical nephrectomy based on tumor stage, size, grade and necrosis: The SSIGN score. J Urol, 168(6), 2395–2400.

[42] Zastrow, S., et al. (2015). Decision curve analysis and external validation of the postoperative Karakiewicz nomogram for renal cell carcinoma based on a large single-center study cohort. World J Urol, 33(3), 381–388.

[43] Jiang, G., Chen, S., & Chen, M. (2020). Exploration of IMDC model in patients with metastatic renal cell carcinoma using targeted agents: A meta-analysis. Int Braz J Urol, 46(3), 328–340.

[44] Motzer, R. J., et al. (2002). Interferon-alfa as a comparative treatment for clinical trials of new therapies against advanced renal cell carcinoma. J Clin Oncol, 20(1), 289–296.

[45] Otto, W. & Drahoslav, L. (1960). Hydrophilic gels in biologic use. Nature, 185, 117–118.

[46] Hennink, W. E. & Van Nostrum, C. F. (2012). Novel crosslinking methods to design hydrogels. Adv Drug Deliv Rev, 64, 223–236.

[47] Rehman, W. U., et al. (2020). Hydrogel: A promising material in pharmaceutics. Curr Pharm Des, 26(45), 5892–5908.

[48] Breedveld, V., et al. (2004). Rheology of block copolypeptide solutions: Hydrogels with tunable properties. Macromolecules, 37(10), 3943–3953.

[49] Guilherme, M., et al. (2003). Hydrogels based on PAAm network with PNIPAAm included: Hydrophilic–hydrophobic transition measured by the partition of Orange II and Methylene Blue in water. Polymer, 44(15), 4213–4219.

[50] Ahmad, Z., et al. (2022). Versatility of hydrogels: From synthetic strategies, classification, and properties to biomedical applications. Gels, 8(3), 167.

[51] Sri, B., Ashok, V., & Arkendu, C. (2012). As a review on hydrogels as drug delivery in the pharmaceutical field. Int J Pharm Chem Sci, 1(2), 642–661.

[52] Ahmed, E. M. (2015). Hydrogel: Preparation, characterization, and applications: A review. J Adv Res, 6(2), 105–121.

[53] Kopeček, J. (2007). Hydrogel biomaterials: A smart future?. Biomaterials, 28(34), 5185–5192.

[54] Oyen, M. (2014). Mechanical characterisation of hydrogel materials. Int Mater Rev, 59(1), 44–59.

[55] Thang, N. H., Chien, T. B., & Cuong, D. X. (2023). Polymer-based hydrogels applied in drug delivery: An overview. Gels, 9(7), 523.

[56] Li, J., et al. (2023). Progress in the treatment of drug-loaded nanomaterials in renal cell carcinoma. Biomed Pharmacother, 167, 115444.

[57] He, Y., et al. (2020). Temperature sensitive hydrogel for preoperative treatment of renal carcinoma. Mater Sci Eng C, 111, 110798.

[58] Wolinsky, J. B., Colson, Y. L., & Grinstaff, M. W. (2012). Local drug delivery strategies for cancer treatment: Gels, nanoparticles, polymeric films, rods, and wafers. J Control Release, 159(1), 14–26.

[59] Dattilo, M., et al. (2023). Polysaccharide-based hydrogels and their application as drug delivery systems in cancer treatment: A review. J Func Biomater, 14(2), 55.

[60] Alallam, B., et al. (2023). Advanced drug delivery systems for renal disorders. Gels, 9(2), 115.

[61] Sood, N., et al. (2016). Stimuli-responsive hydrogels in drug delivery and tissue engineering. Drug Deliv, 23(3), 748–770.

[62] Poustchi, F., et al. (2021). Combination therapy of killing diseases by injectable hydrogels: From concept to medical applications. Adv Healthc Mater, 10(3), 2001571.

[63] Mandal, A., et al. (2020). Hydrogels in the clinic. Bioeng Transl Med, 5(2), e10158.

[64] Ueda, K., et al. (2016). Growth inhibitory effect of an injectable hyaluronic acid–tyramine hydrogels incorporating human natural interferon-α and sorafenib on renal cell carcinoma cells. Acta Biomater, 29, 103–111.

[65] Pishavar, E., et al. (2021). Multifunctional and self-healable intelligent hydrogels for cancer drug delivery and promoting tissue regeneration in vivo. Polymers, 13(16), 2680.

[66] Zhu, T., et al. (2019). Recent progress of polysaccharide-based hydrogel interfaces for wound healing and tissue engineering. Adv Mater Interfaces, 6(17), 1900761.

[67] Vigata, M., et al. (2020). Hydrogels as drug delivery systems: A review of current characterization and evaluation techniques. Pharmaceutics, 12(12), 1188.

[68] Radulescu, D.-M., et al. (2022). New insights of scaffolds based on hydrogels in tissue engineering. Polymers, 14(4), 799.

[69] Ran, P., et al. (2023). Light-triggered theranostic hydrogels for real-time imaging and on-demand photodynamic therapy of skin abscesses. Acta Biomater, 155, 292–303.

[70] Xie, X., et al. (2022). A "sense-and-treat" hydrogel for rapid diagnose and photothermal therapy of bacterial infection. Chem Eng J, 443, 136437.

[71] Mondlane, E. R., et al. (2021). The role of immunotherapy in advanced renal cell carcinoma. Int Braz J Urol, 47, 1228–1242.

[72] Erfani, A., Diaz, A. E., & Doyle, P. S. (2023). Hydrogel-enabled, local administration and combinatorial delivery of immunotherapies for cancer treatment. Mater Today.

[73] Gu, X., et al. (2022). Hydrogel and nanoparticle carriers for kidney disease therapy: Trends and recent advancements. Prog Biomed Eng, 4(2), 022006.

[74] Mohamed, S. M. D. S., Welsh, G. I., & Roy, I. (2023). Renal tissue engineering for regenerative medicine using polymers and hydrogels. Biomater Sci.

[75] Zhao, J., et al. (2022). Progress of research in in situ smart hydrogels for local antitumor therapy: A review. Pharmaceutics, 14(10), 2028.

[76] Kong, F., Mehwish, N., & Lee, B. H. (2022). Emerging albumin hydrogels as personalized biomaterials. Acta Biomater.

[77] Godau, B., et al. (2023). A drug-eluting injectable nanogel for localized delivery of anticancer drugs to solid tumors. Pharmaceutics, 15(9), 2255.

[78] Li, J. & Mooney, D. J. (2016). Designing hydrogels for controlled drug delivery. Nat Rev Mater, 1(12), 1–17.

[79] Kass, L. E. & Nguyen, J. (2022). Nanocarrier-hydrogel composite delivery systems for precision drug release. Wiley Interdiscip Rev Nanomed Nanobiotechnol, 14(2), e1756.

[80] Kumar, S. & Bajaj, A. (2020). Advances in self-assembled injectable hydrogels for cancer therapy. Biomater Sci, 8(8), 2055–2073.

[81] Patil, S. B., et al. (2020). Functionally tailored electro-sensitive poly (acrylamide)-g-pectin copolymer hydrogel for transdermal drug delivery application: Synthesis, characterization, in-vitro and ex-vivo evaluation. Drug Deliv Lett, 10(3), 185–196.

[82] Xiao, T., et al. (2022). Injectable alginate hydrogels for synergistic tumor combination therapy through repolarization of tumor-associated macrophages. J Control Release, 348, 239–249.

Index

https://doi.org/10.1515/9783111334080-009

Printed in the USA
CPSIA information can be obtained
at www.ICGtesting.com
JSHW050017170524
63288JS00010B/175